Advances in
Insect Physiology

Volume 42

Advances in Insect Physiology

Small RNAs: Their Diversity, Roles and Practical Uses

edited by

Elizabeth L. Jockusch

Department of Ecology and Evolutionary Biology
University of Connecticut
Storrs, CT, USA

Volume 42

ELSEVIER

AMSTERDAM • BOSTON • HEIDELBERG • LONDON • NEW YORK • OXFORD
PARIS • SAN DIEGO • SAN FRANCISCO • SINGAPORE • SYDNEY • TOKYO
Academic Press is an imprint of Elsevier

ACADEMIC PRESS

Academic Press is an imprint of Elsevier
32 Jamestown Road, London NW1 7BY, UK
Linacre House, Jordan Hill, Oxford OX2 8DP, UK
Radarweg 29, PO Box 211, 1000 AE Amsterdam, The Netherlands
225 Wyman Street, Waltham, MA 02451, USA
525 B Street, Suite 1900, San Diego, CA 92101-4495, USA

First edition 2012

ISBN: 978-0-12-387680-5
ISSN: 0065-2806

Printed and bound by CPI Group (UK) Ltd, Croydon, CR0 4YY

Transferred to Digital Print 2012

Contents

Contributors

Steven W. Adamson

Department of Biological Sciences, University of Southern Mississippi, Hattiesburg, Mississippi, USA

Xiao-Ya Chen

National Key Laboratory of Plant Molecular Genetics, Institute of Plant Physiology and Ecology, Shanghai Institutes for Biological Sciences, Chinese Academy of Sciences, Shanghai, PR China

Darron A. Cullen[1]

School of Biological Sciences, The University of Sydney, Sydney, Australia

Zhaojun Han

Department of Entomology, College of Plant Protection, Nanjing Agricultural University, Nanjing, China

Yong-Ping Huang

CAS Key Laboratory of Entomology, Institute of Plant Physiology and Ecology, Shanghai Institutes for Biological Sciences, Chinese Academy of Sciences, Shanghai, PR China

Xinglai Ji

Laboratory for Conservation and Utilization of Bio-Resources & Key Laboratory for Microbial Resources of the Ministry of Education, Yunnan University, Kunming, China

Shahid Karim

Department of Biological Sciences, University of Southern Mississippi, Hattiesburg, Mississippi, USA

Fei Li

Department of Entomology, College of Plant Protection, and Key Laboratory of Integrated Management of Crop Diseases and Pests, Nanjing Agricultural University, Ministry of Education, Nanjing, China

[1] Current address: Department of Zoology, University of Cambridge, Cambridge, United Kingdom.

Guoqing Li
Department of Entomology, College of Plant Protection, Nanjing Agricultural University, Nanjing, China

Fei Ma
Laboratory of Comparative Genomics and Bioinformatics & Jiangsu Key Laboratory for Biodiversity and Biotechnology, College of Life Science, Nanjing Normal University, Nanjing, China

Ying-Bo Mao
National Key Laboratory of Plant Molecular Genetics, Institute of Plant Physiology and Ecology, Shanghai Institutes for Biological Sciences, Chinese Academy of Sciences, Shanghai, PR China

Maria-Carla Saleh
Institut Pasteur, Viruses and RNA interference, CNRS URA3015, Paris, France

Yang Sun
Department of Entomology, College of Plant Protection, Nanjing Agricultural University, Nanjing, China

Xiao-Yuan Tao
National Key Laboratory of Plant Molecular Genetics, Institute of Plant Physiology and Ecology, Shanghai Institutes for Biological Sciences, Chinese Academy of Sciences, Shanghai, PR China

Nicolas Vodovar
Institut Pasteur, Viruses and RNA interference, CNRS URA3015, Paris, France

Shuping Wang
Department of Entomology, College of Plant Protection, Nanjing Agricultural University, Nanjing, China

Xue-Yi Xue
National Key Laboratory of Plant Molecular Genetics, Institute of Plant Physiology and Ecology, Shanghai Institutes for Biological Sciences, Chinese Academy of Sciences, Shanghai, PR China

Lun Yu
Department of Entomology, College of Plant Protection, Nanjing Agricultural University, Nanjing, China

Preface

Small RNAs are a diverse, and until recently unappreciated, set of non-protein-coding molecules that function in myriad ways from defence against viruses and transposable elements to regulation of development and physiology, to genome remodeling. At least one type of small RNA was already present in the common ancestor of all eukaryotes, indicating that small RNAs have been diversifying for more than two billion years.

Originally, research on small RNAs progressed along multiple separate lines. MicroRNAs (miRNAs) are now known to regulate many aspects of development in both metazoans and plants. miRNAs were first characterized in 1993 in the nematode worm *Caenorhabditis elegans*. A mutation that disrupted developmental timing did not map to a protein-coding region but rather to an intron which encoded two small RNAs (Lee *et al.*, 1993). Developmental regulation by miRNAs was initially regarded as a peculiarity of worms, and even in worms, the discovery of a second miRNA, *let-7*, took 7 years (Reinhart *et al.*, 2000). However, *let-7* was found to be conserved across bilaterian animals, including the fruit fly *Drosophila melanogaster* (Pasquinelli *et al.*, 2000). This finding and the subsequent characterization of numerous additional miRNAs in *D. melanogaster* (Lagos-Quintana *et al.*, 2001) opened up the study of miRNAs in arthropods.

A second avenue of research also originated in *C. elegans*. In 1998, Fire, Mello and colleagues showed that double-stranded RNA was a potent inducer of targeted gene silencing (Fire *et al.*, 1998). This phenomenon is called RNA interference (RNAi). The evolutionary antiquity of the RNAi mechanism was discovered almost immediately: its effectiveness in *D. melanogaster* was demonstrated within the year (Kennerdell and Carthew, 1998; Misquitta and Paterson, 1999). The discovery of RNAi made possible functional studies of diverse set of organisms and questions, several of which are highlighted in this volume.

These disparate lines of research coalesced with the discovery that the enzyme Dicer is used in processing both types of small RNAs (Hutvágner *et al.*, 2001). Since then, research into small RNAs has taken off in many directions. Although the field is only about 15 years old, we now have a general understanding of the biogenesis, function, origin, diversity, and evolution of an ever-increasing set of small non-coding RNAs in an increasing diversity of species. Arthropods have served as a system for deciphering the mechanisms, functions, and evolution of small RNAs, and RNAi has also provided an important tool for the study of arthropods.

One early contribution of work in *D. melanogaster* was the demonstration that RNAi served an antiviral defense function in animal cells (Li *et al.*, 2002), as it does in plants. This volume opens with a chapter by Nicolas Vodovar and Maria-Carla Saleh exploring this presumed ancestral function of small RNAs and the resulting arms race between viruses and their insect hosts. Because the RNAi mechanism is responsive to exogenous double-stranded RNA, it also forms the basis of our ability to target any gene of interest by introducing double-stranded RNA matching it.

The volume then turns to the uses of small RNAs as tools to study arthropods. The second chapter, by Darron A. Cullen, synthesizes what has been learned about the biology of hemimetabolous insects using an RNAi approach. Despite their important phylogenetic positions, tools for analyses of gene function were lacking in these taxa. Initially used primarily to study gene function during development, RNAi is now also used to study behavior and physiology, and as a tool to manipulate phenotypes of ecological or evolutionary interest.

The potential of RNAi as a mechanism to control insect pests has also received considerable attention, beginning with the demonstration that insect herbivores were affected by double-stranded RNA obtained from their food plants (Baum *et al.*, 2007; Mao *et al.*, 2007). In the third chapter, Xiao-Ya Chen and colleagues consider the prospects, promises, and challenges for this novel method of pest control.

The fourth chapter, by Shahid Karim and Steven W. Adamson, is on the contributions of RNAi to our understanding of tick biology. Ticks are important ectoparasites of vertebrates, including humans. Their unique blood-feeding strategy results in transmission of a large number of disease-causing agents to their vertebrate hosts. RNAi is being used to characterize the function of novel genes involved in blood feeding and to seek molecular targets for regulating both ticks and tick-borne pathogens.

The volume concludes with a chapter by Fei Li and colleagues on methods for identifying small RNAs and their regulatory targets. The small size of the functional component, short regions of perfect complementarity, and our still incomplete understanding of matching rules make this a challenging area, but one in which the combined power of next-generation sequencing and computational approaches has led to rapid progress.

References

Baum, J. A., Bogaert, T., Clinton, W., Heck, G. R., Feldmann, P., Ilagan, O., Johnson, S., Plaetinck, G., Munyikwa, T., Pleau, M., Vaughn, T. and Roberts, J. (2007). Control of coleopteran insect pests through RNA interference. *Nat. Biotechnol.* **25**, 1322–1326.

Fire, A., Xu, S., Montgomery, M. K., Kostas, S. A., Driver, S. E. and Mello, C. C. (1998). Potent and specific genetic interference by double-stranded RNA in *Caenorhabditis elegans*. *Nature* **391**, 806–811.

Hutvágner, G., McLachlan, J., Pasquinelli, A. E., Bálint, E., Tuschl, T. and Zamore, P. D. (2001). A cellular function for the RNA-interference enzyme Dicer in the maturation of the *let-7* small temporal RNA. *Science* **293**, 834–838.

Kennerdell, J. R. and Carthew, R. W. (1998). Use of dsRNA-mediated genetic interference to demonstrate that *frizzled* and *frizzled-2* act in the wingless pathway. *Cell* **95**, 1017–1026.

Lagos-Quintana, M., Rauhut, R., Lendeckel, W. and Tuschl, T. (2001). Identification of novel genes coding for small expressed RNAs. *Science* **294**, 853–858.

Lee, R. C., Feinbaum, R. L. and Ambros, V. (1993). The C. elegans heterochronic gene *lin-4* encodes small RNAs with antisense complementarity to *lin-14*. *Cell* **75**, 843–854.

Li, H., Li, W. X. and Ding, S. W. (2002). Induction and suppression of RNA silencing by an animal virus. *Science* **296**, 1319–1321.

Mao, Y. B., Cai, W. J., Wang, J. W., Hong, G. J., Tao, X. Y., Wang, L. J., Huang, Y. P. and Chen, X. Y. (2007). Silencing a cotton bollworm P450 monooxygenase gene by plant-mediated RNAi impairs larval tolerance of gossypol. *Nat. Biotechnol.* **25**, 1307–1313.

Misquitta, L. and Paterson, B. M. (1999). Targeted disruption of gene function in *Drosophila* by RNA interference (RNA-i): A role for *nautilus* in embryonic somatic muscle formation. *Proc. Natl. Acad. Sci. USA* **96**, 1451–1456.

Pasquinelli, A. E., Reinhart, B. J., Slack, F., Martindale, M. Q., Kuroda, M. I., Maller, B., Hayward, D. C., Ball, E. E., Degnan, B., Müller, P., Spring, J., Srinivasan, A., *et al.* (2000). Conservation of the sequence and temporal expression of *let-7* heterochronic regulatory RNA. *Nature* **408**, 86–89.

Reinhart, B. J., Slack, F. J., Basson, M., Pasquinelli, A. E., Bettinger, J. C., Rougvie, A. E., Horvitz, H. R. and Ruvkun, G. (2000). The 21-nucleotide *let-7* RNA regulates developmental timing in *Caenorhabditis elegans*. *Nature* **403**, 901–906.

Elizabeth L. Jockusch

Of Insects and Viruses: The Role of Small RNAs in Insect Defence

Nicolas Vodovar and Maria-Carla Saleh

Institut Pasteur, Viruses and RNA interference, CNRS URA3015, Paris, France

Abstract

In the past decade, small RNA pathways have been identified as a major mechanism of gene regulation. From an immunity standpoint, these pathways play a central role either by regulating immune reactions or by acting as immune effectors. In insects, several studies have unravelled the role of RNA interference (RNAi) as an antiviral response and have uncovered a complex relationship between insects and viruses that co-evolve in an ongoing race for supremacy. In this review, we comment on the role of small RNA pathways in insect defence and the exploitation of these same pathways by pathogens. We illustrate the host–pathogen relationship under RNAi constraints using several examples and we discuss future directions in using RNAi as a tool to control insect immunity.

ADVANCES IN INSECT PHYSIOLOGY VOL. 42
ISBN 978-0-12-387680-5
DOI: 10.1016/B978-0-12-387680-5.00001-X

1 Introduction

Host–pathogen interactions can be pictured as an arms race between two adversaries. On the one hand, the pathogen deploys virulence factors to exploit the resources of the host. On the other hand, the host fights back with immune responses to clear the pathogen or at least minimise its deleterious effects. The outcome of this interaction can vary from commensalism to the death of one of the two players, depending on the relative strength of the effectors involved and possible escape mechanisms to limit these effects. There are two types of immune responses, relying on different mechanisms and effectors: innate and adaptive. While the latter is specific to vertebrates, innate immunity is present in all multicellular organisms, including insects. In vertebrates, the activation of the innate immune response is a prerequisite to the activation of adaptive immunity. Two theories have been proposed to model the induction of the innate immune response. The first theory relies on the recognition by Pattern Recognition Receptors (PRRs) of invariant molecular patterns (Pathogen-Associated Molecular Pattern—PAMP) that are present in most, if not all, microbes (Janeway, 1989, 1992). According to the second theory, the 'theory of danger', the immune system is elicited by alarm signals sent by injured cells (Matzinger, 1994). These two theories are not mutually exclusive as illustrated by the dual activation of *Drosophila* immune response to both glucans (PAMP) and virulence factors (alarm signal) upon infection by the entomopathogenic fungi *Beauveria bassiana* and *Metarhizium anisopliae* (Gottar *et al.*, 2006). Recently, Polly Matzinger reformulated her initial theory of danger, unifying the recognition by PRRs and the recognition of alarm signals into a single model in which the immune system responds to 'danger signals' from different origins and sources (Matzinger, 2007).

1.1 Insects and viruses

Insects, like other organisms, are subject to infection by viruses with RNA or DNA genomes of different structure and polarity. These viruses fall into two distinct classes, depending on the type of host they use during their replication cycle. Viruses from the first class are transmitted from insect to insect, although their host range may not be solely restricted to insects. Among these viruses, some present a strong agronomical impact by affecting domesticated insects. For instance, the Israeli Acute Paralysis Virus (IAPV) is among the pathogens potentially responsible for the honeybee (*Apis mellifera*) colony collapse disorder, which results in the destruction of the hives (Cox-Foster *et al.*, 2007). Others, like *Densovirus* (Parvoviridae) have been successfully used to eliminate *Galleria mellonella* (greater wax moth) infestations in beehives (Lavie *et al.*, 1965), although the use of viruses for biological control of pests remains limited because of their potential impact on human health.

Viruses of the second class are the arthropod-borne viruses (arboviruses). These viruses have the particularity of alternating between hematophagic invertebrate and vertebrate hosts in an obligate fashion. Several insects are responsible for arbovirus transmission to human or cattle, including the mosquitoes *Aedes* spp. (e.g. Rift Valley Fever Virus, Chikungunya Virus, Dengue Virus, Yellow Fever Virus), *Anopheles* spp. (e.g. O'nyong'nyong Virus) and *Culex* spp. (e.g. Rift Valley Fever Virus, Japanese Encephalitis Virus, West Nile Virus); the sand fly *Phlebotomus* spp. (e.g. Vesicular Stomatitis Virus, VSV) and biting midges in the genus *Culicoides* (e.g. Bluetongue Virus; for review see Mellor, 2000 and references therein). Importantly, arboviral infections are asymptomatic in insects but responsible for severe incapacitating diseases in mammalian hosts, especially humans, suggesting a complex co-evolutionary process. There is no specific treatment against such diseases and vaccines are available for only two of them (the Yellow Fever and Japanese Encephalitis viruses). The strong impact of arboviruses on human health and economy, the spread of these diseases around the globe due to climate changes and travelling habits, and the emergence of new arboviral diseases make clear the importance of finding new strategies to limit arboviral transmission to mammals. More generally, insects play important roles in human life, both beneficial and detrimental; this prompts a desire for better understanding of their immune system in order to protect those that are valuable to humans and limit those that bring adverse effects.

1.2 *DROSOPHILA* AS A MODEL TO STUDY HOST–PATHOGEN INTERACTIONS

Upon viral infection, insects mount a distinctive immune response whose hallmark is RNA interference (RNAi). This defence mechanism was originally identified as an antiviral defence in plants (Ratcliff *et al.*, 1997). In insects, most of our knowledge about RNAi and more generally about small RNA pathways come from studies performed in the fruit fly *Drosophila melanogaster*, which celebrated its hundredth anniversary as a model organism in 2010. Since its introduction in the laboratory by Thomas Morgan to study development, *Drosophila* has been the source of invaluable contributions to genetics, developmental biology, neuroscience and immunology. For instance, the first mutation described by Morgan in 1910 led to the discovery of sex-linkage, introduced genetics as a science and genetic analysis as a powerful tool in biology. Throughout the century, *Drosophila* has benefited from a dynamic and collaborative community that rendered this small fly one of the most accomplished multicellular animal models in which to carry out genetic analysis. In addition, the sequence of its genome showed that more than 60% of the genes involved in human genetic diseases are conserved in *Drosophila* (Bier, 2005). Thus, its affordability, short generation time and amenability to both direct and reverse genetics makes it suitable for intensive analysis and provides an appropriate alternative to vertebrate models when characterizing biological processes.

The major breakthrough in immunology that can be attributed to *Drosophila* is the identification of the Toll (Lemaitre, 2004; Lemaitre *et al.*, 1995b, 1996) and Imd (Lemaitre *et al.*, 1995a) pathways which direct an immune response capable of discriminating and taking the appropriate action against an invading microbe, depending on its class (Lemaitre *et al.*, 1997). The identification of Toll pioneered the discovery of the Toll-like receptors (TLR) in mammals and the subsequent understanding of the mechanisms that govern innate immunity (Rock *et al.*, 1998). More recently, the identification of the RNAi pathway as the major antiviral defence mechanism positioned *Drosophila* as a central model to study insect antiviral immunity (Galiana-Arnoux *et al.*, 2006; van Rij *et al.*, 2006; Wang *et al.*, 2006; Zambon *et al.*, 2006). In the case of arboviruses, enthusiasm for *Drosophila* as a study model over genuine vectors such as mosquitoes may be dampened by the fact that it is not a vector for any arboviral disease. However, the ability of several arboviruses to replicate in *Drosophila* (e.g. Sindbis Virus (SINV): Xiong *et al.*, 1989; West Nile Virus: Chotkowski *et al.*, 2008; Dengue Virus: Sessions *et al.*, 2009; VSV: Mueller *et al.*, 2010) and the quasi-absence of genetic tools available in other insects make *Drosophila* a very powerful model to study all virus–insect, and more specifically arbovirus–insect, interactions.

2 Generalities about insect defence mechanisms

Insects have developed effective defence mechanisms to protect themselves from infections. These defence mechanisms rely on both physical and chemical barriers that prevent microbes from penetrating the body cavity, along with a wide range of inducible reactions that aim at eradicating invading microorganisms. Below, the main aspects of the antimicrobial and antiviral responses in *Drosophila* are summarised; these are likely to be conserved in other insects.

2.1 THE ANTIMICROBIAL IMMUNE RESPONSE

First of all, the cuticle, which covers the insect body, prevents the entry of microbes into the body cavity through the epidermis. The epithelia of the intestinal and respiratory tracts (trachea) are also lined by chitinous membranes that avert direct contact between cells and microbes. In the gut, which constitutes the main route of infection, the secretion of digestive enzymes, a low pH and the production of reactive oxygen species maintain an environment hostile to microbial survival (Lemaitre and Hoffmann, 2007; Tzou *et al.*, 2002). When microbes overcome this first line of defence, the tracheal and intestinal cells secrete antimicrobial peptides (AMPs) to eradicate microbes that would persist and colonise the lumen of their respective epithelia (Tzou *et al.*, 2000).

Once these physical and chemical barriers are breached, the entry of microbes within the body cavity triggers immune reactions with cellular and humoral components. The cellular response mainly involves macrophages, called plasmatocytes, that engulf microbes present in the circulatory system or hemolymph through phagocytosis (for review see Lemaitre and Hoffmann, 2007). Microbes also induce a humoral response whose hallmark is the synthesis and secretion of several AMPs by the fat body. The expression of AMPs is regulated by two parallel NF-κB signalling cascades, the Toll and Imd pathways, whose intracellular components display striking similarity with those of the vertebrate TNF-R (Tumour Necrosis Factor Receptor) and TLR/IL-12R (Interleukin 12 Receptor) pathways, respectively (Lemaitre and Hoffmann, 2007). The Imd pathway is activated predominantly by Gram-negative bacteria and in turn, activates the expression of antibacterial peptide-encoding genes (e.g. *diptericin*), whereas the Toll pathway is predominantly activated by Gram-positive bacteria and fungi, and regulates the expression of genes that encode antifungal peptides (e.g. *drosomycin*) and a subset of antibacterial peptides (Lemaitre and Hoffmann, 2007; Lemaitre *et al.*, 1997). These two pathways are discriminatively activated through the recognition of a small repertoire of PAMPs which consist of a peptidoglycan whose structure differs between Gram-positive and Gram-negative bacteria (Leulier *et al.*, 2003), and the fungal beta-1,3 glucans (Gottar *et al.*, 2006). Although the two pathways are down-regulated to limit the adverse effects of prolonged activation (Zaidman-Remy *et al.*, 2006), AMP concentrations are sustained at high levels over several days, protecting the flies against a second challenge (Boman *et al.*, 1972).

2.2 THE ANTIVIRAL RESPONSE

The core antiviral response of *Drosophila* is distinct from its antimicrobial counterpart. It involves different components that can be divided into two classes depending on the mechanisms by which they are elicited. The first class is directly triggered by the presence of double-stranded RNA (dsRNA) and includes the degradation of the viral genome by the RNAi machinery (Galiana-Arnoux *et al.*, 2006; van Rij *et al.*, 2006; Wang *et al.*, 2006; Zambon *et al.*, 2006) and the regulation of genes like *Vago* via an RNAi-independent function of Dicer-2 (Dcr-2; Deddouche *et al.*, 2008). The second class is more likely to respond to cell damages and involves the Jak/STAT pathway (Dostert *et al.*, 2005) and potentially, the Toll and Imd pathways (Avadhanula *et al.*, 2009; Costa *et al.*, 2009). The RNAi pathways will be extensively discussed in the Section 3; below the role of other pathways in the insect antiviral response is summarised.

The Jak/STAT signalling cascade is involved in numerous developmental processes (for review see Arbouzova and Zeidler, 2006). This pathway is constituted in *Drosophila* by cytokines of the Unpaired family, the transmembrane receptor domeless (dome), the Janus Kinase hopscotch (hop) and the transcriptional activator STAT92. The Jak/STAT pathway was originally thought to be involved in the antimicrobial response (Agaisse *et al.*, 2003); although mutations impairing the pathway do not affect AMP expression or fly survival after microbial infection (Dostert *et al.*, 2005), while *hop*-deficient flies display a strong susceptibility to *Drosophila* C virus (DCV) challenge. A similar antiviral role was proposed for the Jak/STAT pathway in the mosquito *Aedes aegypti* upon dengue infection (Souza-Neto *et al.*, 2009). In *Drosophila*, two lines of evidence suggest that the activation of the Jak/STAT pathway constitutes a secondary response to viral infection, possibly in response to cellular damage: (i) the expression of one of its virus-induced targets, *vir-1*, is upregulated in tissues in which virus is undetectable (Dostert *et al.*, 2005) and (ii) vir-1 is only expressed when flies are infected with live virus but not after injection of UV-inactivated virus or viral dsRNA (Hedges and Johnson, 2008). These latter results correlate with the inability of a constitutively active allele of *hop*, *Tum-l*, to activate *vir-1* expression in the absence of viral infection (Dostert *et al.*, 2005). Taken together, these data suggest that the Jak/STAT pathway responds to cytokine signalling thereby informing cells of viral infection. This role is reminiscent of that of the Jak/STAT pathway in mammals, illustrating the conservation of innate immunity regulatory mechanisms through evolution.

Recently, the antimicrobial pathways Toll and Imd were reported to play an antiviral role in *Drosophila* (Avadhanula *et al.*, 2009; Costa *et al.*, 2009) and the mosquito *A. aegypti* (Xi *et al.*, 2008). However, the reported results appear contradictory possibly because they were obtained from different host-virus interaction models. Because the role of the Toll and Imd pathways in antiviral response is beyond the scope of this review, we will not further discuss these results.

3 RNAi and the immune response

RNAi refers to sequence-specific RNA-dependent silencing mechanisms (Fire *et al.*, 1998; Ratcliff *et al.*, 1997) that regulate various processes such as gene expression, epigenetic modifications and defence against pathogens. These mechanisms are conserved throughout evolution and functional RNAi pathways have been identified in fungi, plants and animals albeit with different roles. Almost all RNAi pathways are triggered by dsRNA that varies in length and origin. The dsRNA molecules are rapidly converted by RNaseIII enzymes called Dicers into small RNAs whose sizes range from 21 to 32 nucleotides, depending on the pathway from which they originate. Next, the small RNAs are loaded into a multiprotein RNA-Induced Silencing Complex (RISC) where they guide the recognition of target RNA through an Argonaute (AGO)/Piwi family member. These AGO/Piwi proteins constitute the heart of the RNAi system, as they bind small RNAs and directly mediate silencing at the transcriptional and post-transcriptional levels.

3.1 THE SMALL RNA PATHWAYS

There are three classes of small RNAs that differ in size, the template from which they originate and the pathway through which they are processed (Fig. 1).

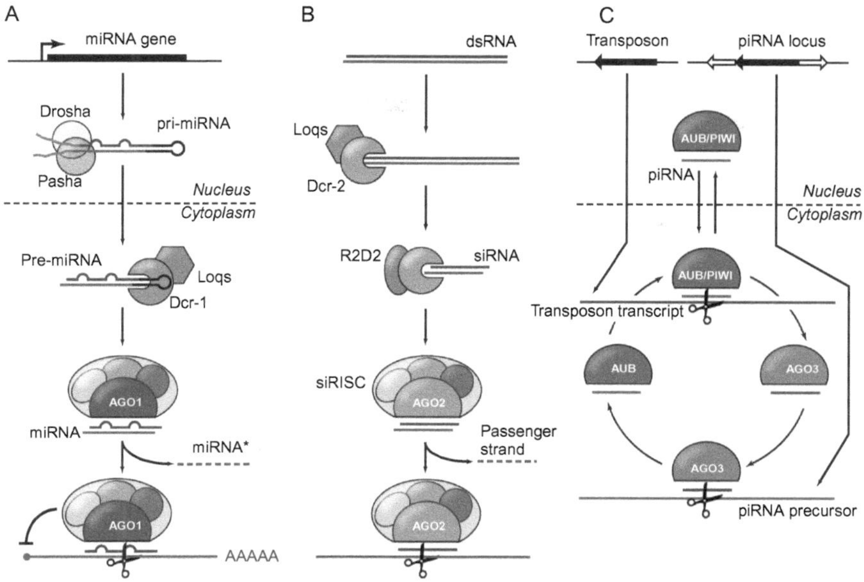

FIG. 1 Schematic representation of the small RNA pathways in *Drosophila*. (A) The miRNA pathway is initiated by the transcription of miRNA genes. Primary miRNA (pri-miRNA) transcripts are first processed by the Drosha/Pasha complex in the nucleus then exported to the cytoplasm as pre-miRNAs. There, Dcr-1 and loquacious (loqs) complete the processing and deliver the mature miRNA to AGO1-containing RISC. The miRNA* (star) is eliminated and the guide miRNA directs translational repression or cleavage of the cognate mRNA. (B) The siRNA pathway is initiated by double-stranded RNA of viral or genomic origin which is recognised and cleaved by Dcr-2 with the help of loqs. The resulting double-stranded siRNAs are delivered to AGO2-containing RISC by Dcr-2 and R2D2. The passenger strand is eliminated and the guide siRNA directs the degradation of the target RNA via AGO2 catalytic activity. (C) The piRNA pathway is initiated by maternally deposited piRNAs loaded into Aubergine (Aub)/Piwi proteins. Through an amplification loop that involves sense and antisense transcripts, AGO3 and Aub, the piRNA pool is amplified. The newly produced piRNAs loaded into Aub are transported to the nucleus where they are thought to be involved in chromatin modifications. (See Color Plate 1 at the back of the book.)

microRNAs (miRNAs; Fig. 1A) are 22 nucleotide-long small RNAs originating from longer genome-encoded precursors that contain an imperfectly base-paired hairpin. This hairpin is first processed by the Drosha/Pasha complex in the nucleus before being exported to the cytoplasm (Denli *et al.*, 2004; Han *et al.*, 2004; Lee *et al.*, 2003). Then, Dicer-1 (Dcr-1) completes processing into miRNAs before loading them into AGO1-containing RISC where they guide translational repression and mRNA degradation through different mechanisms (for review Eulalio *et al.*, 2008). miRNAs constitute the most evolutionarily conserved class of small RNAs and play a critical role in development and the regulation of cellular processes.

Small interfering RNAs (siRNAs; Fig. 1B) are 21 nucleotide-long small RNAs that originate from perfectly base-paired long dsRNA molecules that are processed through a Dcr-2/AGO2 pathway. The siRNA pathway is mostly involved in defence against parasitic nucleic acid elements, such as retrotransposons or viruses. Two subclasses of siRNA can be distinguished based on the origin of the dsRNA. endo-siRNAs are produced from genome-encoded invert-repeated structures or sites of convergent transcription and antisense transcripts from various loci, including retrotransposons and pseudogenes. In *Drosophila*, endo-siRNAs are produced both in the soma and the gonads and are thought to be involved in genome stability by repressing transposition (Brennecke *et al.*, 2007; Chung *et al.*, 2008; Ghildiyal, *et al.*, 2008) and acting on heterochromatin maintenance (Fagegaltier *et al.*, 2009). Although the canonical Dcr-2/AGO2 pathway has not been yet identified in vertebrates, endo-siRNAs were detected in mouse oocytes where they are thought to participate in the control of retrotransposons and pseudogenes (Tam *et al.*, 2008; Watanabe *et al.*, 2008). vsiRNAs are virus-derived siRNAs that originate from viral genomes (dsRNA viruses), viral intermediates of replication (RNA viruses with the exception of retroviruses), or convergent overlapping transcriptional units (DNA viruses). vsiRNAs play a critical role in the antiviral response in insects, nematodes and plants, but to date, this role has not been shown to be conserved in vertebrates (Umbach and Cullen, 2009).

Piwi-interacting RNAs (piRNAs; Fig. 1C) are longer, small RNAs (24–32 nucleotides) and differ from the miRNAs and siRNAs in three main features. First, piRNAs are believed to be processed from single-stranded primary transcripts that are transcribed from defined genomic regions rather than dsRNA molecules (Aravin *et al.*, 2007a). Second, although their biogenesis is not completely understood, it only requires Piwi proteins and seems to be independent of Dicer (Houwing *et al.*, 2007; Vagin *et al.*, 2006). Third, they contribute to the silencing of transposable elements exclusively in the animal gonads (Aravin *et al.*, 2003, 2007b; Brennecke *et al.*, 2007; Olivieri *et al.*, 2010; Vagin *et al.*, 2006).

3.2 ANTIVIRAL RNAI IN INSECTS

RNAi was first identified as an antiviral defence mechanism in plants (Ratcliff *et al.*, 1997) and in the nematode *Caenorhabditis elegans* (Lu *et al.*, 2005; Schott *et al.*, 2005; Wilkins *et al.*, 2005). In insects, RNA-dependent silencing of viral replication was first described using recombinant SINV-expressing Dengue Virus 2 (DENV2) sequences from a duplicated subgenomic promoter. Mosquitoes infected with the recombinant SINV were resistant to a second challenge by the wild-type DENV2 through an RNA-dependent mechanism (Adelman *et al.*, 2001; Gaines *et al.*, 1996; Olson *et al.*, 1996). Similar results indicating an RNA-based mechanism were observed with recombinant SINV-expressing Rift Valley virus sequences (Billecocq *et al.*, 2000). The link between the above observations and RNAi was established in *Drosophila* where the depletion of *AGO2* led to the accumulation of Flock House virus (FHV) RNAs in cell culture (Li *et al.*, 2002); these observations were later confirmed in a mosquito (Keene *et al.*, 2004). More recently, several studies confirmed the role of RNAi as an antiviral response in insects and the siRNA pathway as a major player in fighting against viral infection. Indeed, flies carrying null alleles of *Dcr-2*, *AGO2* or *R2D2* (see below for details) are more susceptible to viral infection as illustrated by their inability to control viral replication and by their premature death compared to wild-type individuals. Although RNAi has only been shown to exert an antiviral function in *Drosophila* (Galiana-Arnoux *et al.*, 2006; van Rij *et al.*, 2006; Wang *et al.*, 2006; Zambon *et al.*, 2006) and mosquitoes (Campbell *et al.*, 2008; Cirimotich *et al.*, 2009; Keene *et al.*, 2004; Khoo *et al.*, 2010; Sanchez-Vargas *et al.*, 2009), several insects display a functional RNAi machinery presumably also involved in antiviral defence (Table 1).

From an evolutionary standpoint, some genes in the antiviral RNAi pathways are shown to display a high evolutionary rate compared to non-immune genes (Obbard *et al.*, 2006). A similar observation was made about the genes that encode the intracellular components of the antimicrobial pathways (for review Lazzaro, 2008). These data have been interpreted as the result of a constant co-evolution between hosts and pathogens, the latter exerting suppressive strategies on host genes, forcing their adaptation (Lazzaro, 2008). Interestingly, among the strategies developed by viruses to escape from RNAi are the viral suppressors of RNAi (VSRs; for further details see Section 4.1). These VSRs affect various steps in the RNAi pathways and may explain the fast evolution of RNAi genes whose products are targeted by VSRs.

3.3 THE SIRNA PATHWAY: THE MAJOR ANTIVIRAL REACTION IN INSECTS

Among small RNA pathways, the siRNA pathway plays a major role in defence against viral infection. This pathway is composed of two core components that display antiviral function: Dcr-2 and AGO2-containing RISC.

3.3.1 *Dicer-2*

Upon viral infection, dsRNA is produced either as the viral genome itself or as an intermediate of replication. The viral dsRNA is first recognised and processed as double-stranded siRNA by Dcr-2 with the help of loquacious (Marques *et al.*, 2010). From a structural point of view, Dcr-2 includes the following domains from its N- to its C-terminus: (i) a DExH-box helicase domain, (ii) two RNAseIII domains responsible for dicing of the dsRNA and (iii) a dsRNA-binding domain (Lee *et al.*, 2004). Dcr-2 plays a fundamental role in antiviral defence through several aspects of its function. First, the cleavage of viral dsRNA by the RNase III activity is in itself an antiviral mechanism (Flynt

TABLE 1

Examples of insects displaying an RNAi response *in vivo* following uptake of exogenous dsRNA[a]

	Silencing occurs in:
Blattaria	
Blattella germanica	Epidermis, thoracic gland and fat body (Martin *et al.*, 2006)[b]
Diploptera punctata	Corpora allata (endocrine gland) (Lungchukiet *et al.*, 2008)
Coleoptera	
Harmonia axyridis	Wing imaginal discs (Ohde *et al.*, 2009)
Monochamus alternatus	Epidermis (Niu *et al.*, 2008)
Sitophilus spp.	Bacteriome tissue (Vallier *et al.*, 2009)
Tenebrio molitor	Hemolymph, presumably hemocytes or fat body (Zhao *et al.*, 2005)
Tribolium castaneum	Progeny (Bucher *et al.*, 2002)
Diptera	
Aedes spp.	Fat body (Bartholomay *et al.*, 2004)
Anopheles gambiae	Fat body (Blandin *et al.*, 2002)
Armigeres subalbatus	Hemocytes (Infanger *et al.*, 2004)
Bactrocera dorsalis	Ovary (Chen *et al.*, 2008)
Ceratitis capitata	Systemic (Vannini *et al.*, 2010)
Culex pipens	Systemic (Sim and Denlinger, 2008)
Drosophila melanogaster	Central nervous system (Dzitoyeva *et al.*, 2001)
Glossina spp.	Fat body (Lehane *et al.*, 2008)
Lutzomyia longipalpis	Systemic (Sant'Anna *et al.*, 2008)
Hemiptera	
Acyrthosiphon pisum	Ubiquitous and gut (Jaubert-Possamai *et al.*, 2007)
Bemisia tabaci	Midgut and salivary glands (Ghanim *et al.*, 2007)
Nilaparvata lugens	Ubiquitous, gut and central nervous system (Liu *et al.*, 2010)
Planococcus citri	Embryo (soaking of embryos, Volpi *et al.*, 2007)

Rhodnius prolixus	Salivary glands (injection and ingestion, Araujo *et al.*, 2006)
Triatoma brasiliensis	Gut (Araujo *et al.*, 2007)
Hymenoptera	
Apis mellifera	Fat body (Amdam *et al.*, 2003)
Nasonia vitripennis	Progeny (Lynch *et al.*, 2006)
Lepidoptera	
Bombyx mori	Silk gland (Tabunoki *et al.*, 2004)
Epiphyas postvittana	Gut (ingestion, Turner *et al.*, 2006)
Helicoverpa armigera	Midgut (Sivakumar *et al.*, 2007)
Hyalophora cecropia	Ovary/embryo (Bettencourt *et al.*, 2002)
Manduca sexta	Fat body and hemoytes (Eleftherianos *et al.*, 2006)
Spodoptera spp.	Midgut (Rajagopal *et al.*, 2002)
Orthoptera	
Gryllus bimaculatus	Systemic (Meyering-Vos *et al.*, 2006)
Locusta migratoria	Progeny (He *et al.*, 2006)
Schistocerca spp.	Eye (Dong and Friedrich, 2005)

[a] This list does not consider experiments involving injection of dsRNA in pre-blastoderm embryos as they do not reflect dsRNA uptake.

[b] Given the extensive literature available on this subject, we have only considered the first publication, to our knowledge, that reported an RNAi response *in vivo*. In all the experiments listed dsRNA was administered by intra-hemocoelic injection at larval or adult stages unless otherwise stated.

et al., 2009). Indeed, the degradation of dsRNA reduces the amount of template available for the production of new genomes or RNA that encode viral proteins. Second, the production of siRNA is absolutely required for subsequent steps in the RNAi pathway including the formation of an active RISC complex (see below). Third, Dcr-2 is involved in the regulation of some antiviral genes including *Vago* (Deddouche *et al.*, 2008). This role of Dcr-2 in regulating, directly or indirectly, an antiviral transcriptional response is proposed to be dependent on the helicase domain of Dcr-2. However, none of the *Dcr-2* alleles used in this study produced siRNA (Lee *et al.*, 2004) and it thus remains to be established whether the virus-dependent induction of *Vago* is regulated by the sole recognition of dsRNA by Dcr-2, its dicing activity or the presence of viral siRNA. Interestingly, as observed with *vir-1*, *Vago* is not upregulated by the sole presence and processing of dsRNA but requires viral replication, suggesting a more complex mechanism of regulation. Together, these data show that Dcr-2 is a PRR, capable of sensing and processing dsRNA as a signature of viral infection and of initiating both an RNAi-dependent and an RNAi-independent response.

3.3.2 *RNA-induced silencing complex*

Once cleaved from a dsRNA precursor by Dcr-2, double-stranded siRNAs are transferred by Dcr-2 to siRISC in an asymmetric manner (Schwarz *et al.*, 2003; Tomari *et al.*, 2004) with the help of R2D2 (Liu *et al.*, 2003, 2006). RISC is a ribonucleoproteic complex that contains several components including AGO2 (Hammond *et al.*, 2001), dFXR (Drosophila ortholog of fragile X mental retardation protein, Caudy *et al.*, 2002), VIG (vasa intronic gene, Caudy *et al.*, 2002) and siRNA, while other components promote siRISC assembly/activation including aubergine (Specchia *et al.*, 2008), C3PO (Liu *et al.*, 2009) or Hsp90 (Miyoshi *et al.*, 2010). Once the double-stranded siRNA is loaded into siRISC, one of the two strands, called the passenger strand, is eliminated through its cleavage by AGO2 (Matranga *et al.*, 2005; Rand *et al.*, 2005). The antiviral activity of the resulting active siRISC resides in its ability to 'search and destroy' target/viral RNAs and to degrade them in a sequence-specific fashion (Liu *et al.*, 2004). The specificity of this mechanism is provided by the perfect base-pairing between the remaining siRNA guide strand and its target which precedes the cleavage of the target RNA by AGO2 at position 10 with respect to the 5′ extremity of the siRNA. Importantly, the guide strand-assisted cleavage activity of AGO2 is strongly dependent on the perfect base-pairing between the siRNA and its target. Indeed, mutations located around position 10 strongly affect AGO2 cleavage activity and it is believed to strongly impair its antiviral activity (Boden *et al.*, 2003; Das *et al.*, 2004; Gitlin *et al.*, 2005; Wilson and Richardson, 2005 and see Section 4.2).

Recently, it was also shown that AGO2 can also repress translation (Iwasaki *et al.*, 2009). Although AGO2 inhibition of translation has not been studied in the context of the antiviral response, it is tempting to think that such a mechanism could also account for AGO2 antiviral activity in insects.

3.3.3 *Other players*

Recently, a link between the siRNA and the miRNA pathways in antiviral response was proposed with the identification of Ars2 and the nuclear proteins CBP20 and CBP80 as components of the small RNA pathways (Sabin *et al.*, 2009). Flies deficient for either gene display an increased susceptibility to viral infection with DCV or VSV. While these genes interact with components of the miRNA (microprocessor) and siRNA (Dcr-2) pathways, the relevance of each pathway in the phenotypes observed and the role of these genes in controlling viral replication independently of the small RNA pathways remain to be established.

3.3.4 *Amplification: Making silence louder*

The mechanism of RNAi amplification is a specific feature of organisms whose genomes encode RNA-dependent RNA polymerases (RDRs[1]) such as plants (Vaistij *et al.*, 2002; Voinnet *et al.*, 1998) and *C. elegans* (Sijen *et al.*, 2001). Amplification was first discovered through a process called transitivity. In transitivity, aberrant ssRNA transgene-derived transcripts are converted by RDR into dsRNA molecules that are processed into siRNA by Dicer proteins. The resulting siRNAs guide the synthesis of dsRNA from an ssRNA template (Alder *et al.*, 2003; Makeyev and Bamford, 2002; Pak and Fire, 2007). These secondary dsRNAs then produce secondary siRNAs, which in the case of transgene-induced silencing, extend over the originally targeted sequence (Fig. 2). So far, the importance of amplification in antiviral defence has only been demonstrated in plants as illustrated by the increased susceptibility of RDR mutants to viral infection (Schwarz *et al.*, 2003; Wang *et al.*, 2010; Xie *et al.*, 2001). Of note, as RDR are mandatory for RNAi in *C. elegans* (Sijen *et al.*, 2001; Smardon *et al.*, 2000), the role of amplification cannot be uncoupled from RNAi itself.

In *Drosophila*, the existence of transitivity was addressed by testing the production of secondary siRNAs directed against Green Fluorescent Protein (GFP) fusion constructs while targeting the GFP sequence by primary siRNAs produced from a genome-encoded hairpin (Roignant *et al.*, 2003). Under these conditions, the authors did not identify secondary siRNA, suggesting the absence of transitivity and thus amplification in *Drosophila*, consistent with the absence of a canonical RDR-encoding gene in this organism (Roignant *et al.*, 2003) and in insects in general. However, the recent identification of an RDR activity fostered by the largest subunit of the *Drosophila* RNA polymerase

[1] Of note, eukaryotic RNA-dependent RNA polymerase is abbreviated as RDR while those of RNA viruses are abbreviated as RdRp.

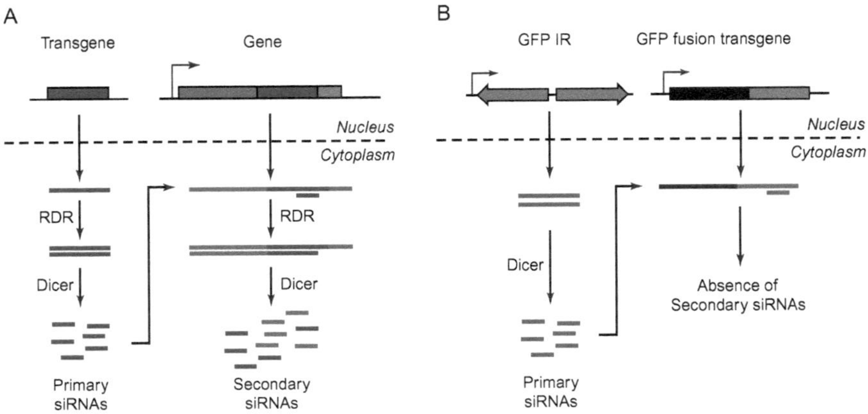

FIG. 2 Examples of transitivity assays in plants (A) and *Drosophila* (B). (A) In plants, aberrant transcripts produced from an integrated transgene corresponding to a fragment of an endogenous gene are converted into dsRNA by an RNA-dependent RNA polymerase (RDR) and cleaved by Dicer. The resulting siRNAs (red) are used to prime the RDR-dependent synthesis of a dsRNA molecule using the homologous endogenous gene transcript as template. Consequently, the newly synthesised dsRNA molecule contains sequence of the surrounding transcript (blue). The presence of secondary siRNAs corresponding to the surrounding regions (blue) is a signature of transitivity. (B) A similar assay was designed in *Drosophila* but failed to identify secondary siRNAs. An inverted repeat (IR) directed against the GFP (green) was expressed from the fly genome. GFP siRNAs were detected (green). However, secondary siRNAs corresponding to the surrounding sequences (black) of a GFP fusion transcript were not detected arguing against transitivity in *Drosophila*. (See Color Plate 2 at the back of the book.)

II elongation factor *in vitro* (D-elp1, Lipardi and Paterson, 2009) reopened the debate on transitivity and amplification in insects. Indeed, several lines of evidence suggest a role of D-elp1 in RNAi: (i) depletion of *D-elp1* in *Drosophila* S2 cells reduces siRNA-mediated gene silencing when cells are soaked with dsRNA, (ii) this depletion is accompanied with an over-representation of transposon-derived transcripts and (iii) immunoprecipitation experiments suggest that D-elp1 interacts with Dcr-2 consistent with a cytoplamic localisation of D-elp1 (Lipardi and Paterson, 2009). Nevertheless, several questions must be addressed, among others: (i) does D-elp1 exhibit an RDR activity *in vivo*?, (ii) is this RDR activity required for D-elp1 function in RNAi? and (iii) in which RNAi-based process (e.g. genome stability through transposition silencing, antiviral defence) is D-elp1 required?

3.3.5 *Systemic RNAi: Sharing the silence*

In plant and worms, the effects of RNAi are not only cell autonomous but spread across the entire organism away from the initial site of silencing (for review Voinnet, 2005). However, the transport of the silencing signal differs drastically in these two systems. In plants, the silencing signal (small RNAs) is spread via plasmodesmal junctions that link almost all the cells together (short-distance spread) and the vascular system (phloem, long-distance spread) while in *C. elegans*, the cell-to-cell spread of the RNAi effect relies on transmembrane transporters of the SID family (Feinberg and Hunter, 2003; Winston *et al.*, 2002, 2007). The spread of RNAi plays a critical role in antiviral defence in plants by initiating an RNAi response against the virus in the entire organism. This protects uninfected cells from infection and thus confines the deleterious effects of viral infection to a limited number of cells. In *Drosophila*, the absence of *sid-1*-related genes and the strict cell autonomy of RNAi effects when initiated from genome-encoded hairpin constructs (Roignant *et al.*, 2003) led to the belief that insects lacked a systemic RNAi pathway. However, the administration of exogenous dsRNA by injection was reported to induce gene knockdown in a growing number of insects from various orders including Diptera, Hymenoptera, Hemiptera, Coleoptera, Lepidoptera and Orthoptera (Table 1). In addition, similarly to *C. elegans* (Timmons and Fire, 1998), some insects are capable of initiating an RNAi response after ingestion of dsRNA, including the honeybee (Patel *et al.*, 2007), the beet armyworm *Spodoptera exigua* (Tian *et al.*, 2009), the brown planthopper *Nilaparvata lugensthe* (Chen *et al.*, 2010), the light brown apple moth *Epiphyas postvittana* (Turner *et al.*, 2006), the tobacco hornworm *Manduca Sexta*, the red flour beetle *Tribolium castaneum* and the pea aphid *Acyrthosiphon pisum* (Whyard *et al.*, 2009). Finally the genomes of several insects including the honey bee, the parasitic wasp *Nasonia vitripennis*, the silkworm *Bombyx mori*, the aphid *Aphis gossypii* and *T. castaneum* encode *sid-1*-like genes, while these genes are absent from other insects including *Drosophila* and mosquitoes (Tomoyasu *et al.*, 2008; Xu and Han, 2008). However, it remains unclear whether these genes play a role in systemic RNAi. Together, these findings show that insect cells are able to take up dsRNA and also suggest that systemic RNAi may occur experimentally in some insects.

More recently, two studies reported further evidence strongly suggesting systemic antiviral RNAi spread upon viral infection *in vivo* in *Drosophila* (Saleh *et al.*, 2009) and *in vitro* in *Aedes albopictus* cells (Attarzadeh-Yazdi *et al.*, 2009). Attarzadeh-Yazdi and colleagues showed spread of an 'RNAi signal' most likely transported from cell-to-cell in mosquito cell culture after infection with Semliki Forest virus (Attarzadeh-Yazdi *et al.*, 2009). In adult *Drosophila*, it was proposed that upon viral infection, cells release an 'RNAi signal' through either cell lysis or membrane shedding that is taken up by surrounding uninfected cells to initiate an immune RNAi antiviral response. Impairing the dsRNA uptake pathway in flies dramatically increased susceptibility to viral infection and inability to control viral replication (Saleh *et al.*, 2009). While the spreading signal remains to be characterised, previous studies *in vitro* show that *Drosophila* hemocyte-derived S2 cells have an active and selective mechanism for uptake of only long dsRNA from the surroundings

(Saleh *et al.*, 2006). Based on the results provided by the studies in *Drosophila*, it appears that the spread signal is likely conveyed through the hemolymph rather than by cell-to-cell transport, although this hypothesis remains to be tested. Further experiments are also required to confirm the type of RNA molecules that are spread *in vivo*, to determine the range of action of systemic RNAi and to identify possible proteins involved in RNA transport.

3.4 OTHER DEFENCE MECHANISMS INVOLVING SMALL RNAS

3.4.1 *Other potential role for siRNAs*

In addition to its direct antiviral activity, the siRNA pathway may be involved in self-regulatory feedback loops aimed at boosting the antiviral response. For instance, in *Drosophila* the *AGO2* and *CG7739* genes overlap at the level of their respective 3′UTRs (Flybase at http://www.flybase.org). Interestingly, small RNA profiling reveals the presence of siRNA produced from a putative dsRNA that would correspond to the base pairing of the 3′UTRs of these two genes (B. Berry and C. Antoniewski, personal communication). A possible interpretation is that upon viral infection, RNAi would be reoriented towards the degradation of the viral genome titrating out endo-siRNAs and therefore reducing the impact of *AGO2/CG7739* repression by endo-siRNAs. As a result, the *AGO2* transcript would be less targeted for degradation and the AGO2 protein level would increase together with the efficiency of the RNAi response. This model is theoretical and requires further investigation.

3.4.2 *The role of miRNAs in insect immune response*

3.4.2.1. Cellular miRNAs and insect defence miRNAs have emerged as an important class of positive and negative regulators involved in multiple facets of both the innate and adaptive immune response in mammals. Indeed, while the miR-17–92 cluster, miR-150, miR-155, miR-181 and miR-223 are involved in the maturation, proliferation and differentiation of T-cells and B-cells, miR-9, miR-146a and miR-155 are believed to negatively regulate acute immune responses through the down-regulation of proteins involved in the receptor-induced signalling pathways (e.g. TLR, for review Tsitsiou and Lindsay, 2009). In insects, little is known about the regulation of immune pathways by miRNA. However, prediction of miRNA targets in *Drosophila* suggests that some immune-related genes, like *Imd*, could be regulated by miRNA (Microcosm at http://www.ebi.ac.uk/enright-srv/microcosm/htdocs/targets/v5/). Nevertheless, experimental validation of this hypothesis is required. In addition, several components of RNAi pathways are suggested as potential miRNA targets in vertebrates (John *et al.*, 2004). In the case of the antiviral response, the possible involvement of miRNA in regulating the siRNA pathway would add another layer of small RNA regulation to this already complex network.

3.4.2.2. Viral-encoded miRNAs Viral-encoded miRNA were first identified in mammalian viruses including Adenovirus, Polyomavirus (e.g. SV40) and Herpesvirus (e.g. Epstein Bar virus, Kaposi's sarcoma-associated herpesvirus or human cytomegalovirus) where they regulate the host and their own gene expression to modulate the host response or control, for instance, the transition between latent and lytic cycles (for review Cullen, 2010). Recently, miRNAs derived from an insect baculovirus were identified and predicted to target both viral and host transcripts (Singh *et al.*, 2010). However, the role of these miRNAs in the physiopathology of baculovirus infection remains to be investigated. Interestingly, viral-encoded miRNAs have only been identified in DNA viruses with a nuclear stage. Although viral miRNAs have been predicted in the genome of RNA viruses (Li *et al.*, 2008), a study by Pfeffer and colleagues failed to identify any miRNAs produced by the Yellow Fever virus or the hepatitis C virus (Pfeffer *et al.*, 2005). Another study identified small RNAs produced in mammalian cells after infection by various RNA viruses (Parameswaran *et al.*, 2010). However, it remains to be established whether these virus-derived small RNAs are siRNAs, miRNAs or degradation products. One can argue that the absence of miRNAs in RNA viruses is due to replication occuring in the cytoplasm, away from Drosha and Pasha which are essential for the biogenesis of host-encoded miRNA. However, it was shown recently that the Epstein Bar virus-encoded miRNA BART2 can be fully processed in the cytoplasm of mammalian cells when expressed by the tick-borne encephalitis virus in a Drosha-independent fashion (Rouha *et al.*, 2010). Thus RNA viruses may support miRNA production and processing but their existence remains to be demonstrated.

3.4.2.3. Viral hijacking of cellular miRNAs It was shown that viruses could exploit cellular miRNAs for their own benefit. For instance, contrary to the seasonal flu viruses, the H1N1 virus responsible for the Spanish flu in 1918 upregulates the expression of miRNAs in the lung that are involved in the regulation of immune reactions and cell death (Li *et al.*, 2010). Along the same line, VSV upregulates the expression of several miRNAs upon infection of mammalian cells. In particular, the VSV-induced over-expression of miR706 inhibits VSV-induced cells death by limiting caspase-3 and -9 activation. Interestingly, as an arbovirus, VSV also infects insects. The potential role of VSV-induced miRNAs in down-regulating insect immunity would provide clues about the evolutionary strategies developed by these viruses as adaptations to disparate vertebrate and invertebrate host.

3.4.2.4. Other possible roles for the miRNA pathway in insect defence The miRNA pathway may directly contribute to the insect antiviral response. Two recent articles reported that siRNAs generated from imperfectly base-paired long dsRNA molecules are preferentially loaded into AGO1 rather than AGO2 (Ghildiyal *et al.*, 2010; Tomari *et al.*, 2007). In *in vitro* conditions, these mismatched siRNAs are thought to act as miRNAs.

In *Dcr-2* mutant flies infected with VSV, a peak of 21 nucleotide-long VSV-derived small RNAs has been observed (Mueller *et al.*, 2010). This result suggests that another dicing activity produces siRNA-like small RNAs in the absence of Dcr-2. This could be achieved by the other known Dicer protein in *Drosophila*, Dcr-1, which has been thought to be dedicated to miRNA biogenesis. Unfortunately, the *Dcr-1* and *AGO1* mutants in *Drosophila* are embryonic lethal and thus non-compliant to any analysis. Nevertheless, it is possible that the absence of Dcr-2 may have revealed a role of Dcr-1 in antiviral response, although this role would be minor compared to that of the siRNA pathway.

3.4.3 *A potential role for piRNA in antiviral defence*

Little is known about the role of piRNA outside the context of controlling transposition in the gonads. Two early reports showed that *piwi*-family mutants (*piwi* and *aubergine*) in *Drosophila* (Zambon *et al.*, 2006) were more sensitive to viral infection and that *Anopheles gambiae* in which *Ago3* had been knocked down displayed increased viral titres when compared to wild-type individuals (Keene *et al.*, 2004). More recently, the production of long viral-derived small RNAs has been reported upon viral infection in *Drosophila* ovarian somatic sheet cells (OSS, Wu *et al.*, 2010) and in the *A. albopictus* C6/36 cell line (Brackney *et al.*, 2010; Scott *et al.*, 2010). While further demonstration will be necessary to assess the nature of these small RNAs in both systems, they share striking features with endogenous piRNAs: (i) they are 25–30 nucleotide long with a size distribution peaking at 27 and 28, (ii) they are strand biased, with most of the reads matching against one strand of the viral genome; this is unlike viral-derived siRNAs and (iii) they display a strong nucleotide bias as previously described for endogenous piRNAs (Brennecke *et al.*, 2007). However, while OSS cells only produced primary piRNA after infection (Wu *et al.*, 2010), C6/36 cells produced primary and secondary piRNAs (N. Vodovar, unpublished observation based on analysis of the data published in Brackney *et al.*, 2010). Although these observations evoke a possible gonadal origin for C6/36 cells, Scott and colleagues proposed a somatic origin for these cells with the lack of Dcr-2 activity compensated by activation of the piRNA pathway (Scott *et al.*, 2010).

Together, these results strongly suggest that the piRNA pathway is triggered and protects against viral infection. Thus the piRNA pathway should be considered as part of the antiviral RNAi response. Of note, C6/36 cells have been reported to silence viral replication in an RNAi-mediated manner (e.g. Adelman *et al.*, 2002), even in the absence of Dcr-2 activity (Scott *et al.*, 2010). Therefore, we would like to stress that contrary to the conclusion of two recent studies (Brackney *et al.*, 2010; Scott *et al.*, 2010) this cell line displays a functional antiviral RNAi response.

3.4.4 *RNAi-based methods to prevent insect infection: were scientists caught off-guard?*

Insects and humans interact in a very complex fashion: some insects are domesticated for products they generate (e.g. honey or silk), some insects are disease vectors (e.g. mosquitoes) and some are agricultural pests (e.g. aphids). Ideally, one would hope to protect useful insects from infection, limit arboviral replication in vectors and control pests, all with fewer adverse effects than chemical pesticides. In the recent years, given the roles of RNAi in gene silencing and antiviral responses in insects, several strategies were proposed to achieve these goals. For instance, feeding pests with dsRNA targeting vital genes was demonstrated to be efficient in laboratory conditions (Whyard *et al.*, 2009) as a proof of principle for the development of insect-specific insecticides.

As mentioned earlier, the IAPV is potentially responsible for colony collapse disorder (CCD, Cox-Foster *et al.*, 2007). It was recently shown that feeding bees with dsRNA directed against the IAPV successfully reduced the effect of an IAPV infection (Maori *et al.*, 2009). However, while this approach is appealing as a way to protect beehives against CCD, it cannot be extended to non-social insects. Along the same line, the immunisation of either *Drosophila* or mosquitoes with dsRNA corresponding to a fragment of a virus, protected the insects against a challenge with the corresponding virus (Keene *et al.*, 2004; Saleh *et al.*, 2009). However, as a general strategy, the individual immunisation of arboviral vectors in the wild by injecting protective dsRNA is obviously an infeasible task. A more practical approach was developed based on direct production of an antiviral dsRNA by the mosquito. This was achieved by generating transgenic *A. gambiae* carrying a fragment of the Dengue 2 virus genome expressed as an inverted repeat (Franz *et al.*, 2006). While such transgenic mosquitoes are efficiently protected against Dengue 2 virus infection over 13 generations, this protective effect is reduced or eventually lost a few generations later through an uncharacterised mechanism (Franz *et al.*, 2006, 2009). This approach is promising although obstacles remain, such as the release of transgenic animals in nature and the displacement of endogenous populations by virus-resistant strains.

Nonetheless, insects did not wait for scientist to develop such antiviral strategies. Indeed, several studies have highlighted the presence of cDNA sequences of non-retroviral RNA viruses in fungi, plant, insects or vertebrates (e.g. Belyi *et al.*, 2010; Crochu *et al.*, 2004; Frank and Wolfe, 2009; Geuking *et al.*, 2009; Horie *et al.*, 2010; Tanne and Sela, 2005; Taylor *et al.*, 2010). Among these studies, one clearly correlates this viral integration with virus protection in insects (Maori *et al.*, 2007). IAPV does not provoke CCD in all the beehives; some are resistant. Recently, it was shown that individuals that are resistant to viral infection carried integrated fragment of IAPV in their own genome. Interestingly, in contrast to bees whose genome is free of IAPV sequence, those that harbour integrated IAPV are not infected by latent IAPV and are resistant to IAPV challenge by a yet uncharacterized mechanism (Maori *et al.*, 2007). Regardless of the mechanisms involved, it appears that the integration of viral sequence by some eukaryotic genomes may constitute a heritable mechanism of protection against some viral diseases adding a new weapon to the arsenal deployed by insects to fight against viruses.

4 Viral escape from RNAi

As part of the arms race between hosts and pathogens, viruses have evolved mechanisms to minimise or suppress the effect of the host response, which aims at limiting viral replication. Given its major role in antiviral defence in insects, RNAi is not an exception to the rule.

4.1 VIRAL SUPPRESSORS OF RNAI

VSRs are viral-encoded proteins that suppress the effect of RNAi. They were first identified in plants (Anandalakshmi *et al.*, 1998; Brigneti *et al.*, 1998; Kasschau and Carrington, 1998; Li *et al.*, 1999) but several insect viruses such as the DCV, the Cricket Paralysis virus (CrPV) and the FHV also encode VSRs. As the role of VSRs has already been extensively reviewed (e.g. Li and Ding, 2006), only their main features are excerpted below.

Although VSRs share the same function, they interfere with the RNAi machinery at different steps through different mechanisms. For instance, (i) DCV1A (DCV) binds to long dsRNA replication intermediates, thus protecting them from being cleaved by Dcr-2 (van Rij *et al.*, 2006), (ii) P19 (plant tombusvirus) sequesters siRNAs thus preventing their loading into AGO2-containing RISC (Silhavy *et al.*, 2002), (iii) CrPV1A (CrPV) directly interacts with AGO2 and compromises its activity (Nayak *et al.*, 2010) and (iv) P20 and CP of the citrus tristeza virus inhibits the systemic spread of RNAi effects (Lu *et al.*, 2004, for a more detailed list of VSRs see Li and Ding, 2006).

Given the major role of dsRNA molecules in RNAi, it is not surprising that most of the VSRs are dsRNA-binding proteins. However, the structures they adopt are not related (Li and Ding, 2006) even though VSRs of viruses belonging to the same family are often present at equivalent genomic locations. Along the same line, the DCV and the CrPV are closely related viruses, as illustrated by the strong similarities between the sequences of their non-structural proteins (71%). However, the sequence and the function of their VSRs are completely unrelated (see above). Together, these observations strongly suggest that the acquisition of VSRs results from convergent evolutionary mechanisms, resulting from selection favouring viruses that escape the effects of RNAi.

Finally, some viruses such as the SINV do not encode any known VSR. Under normal conditions, SINV infection is innocuous for mosquitoes and *Drosophila*. However, when VSRs are expressed from its genome, SINV becomes pathogenic (FHV B2, Cirimotich *et al.*, 2009, DCV1A and CrPV1A, Nayak *et al.*, 2010). Importantly, neither of these suppressors provokes such deleterious effects when over-expressed in *Drosophila* (Berry *et al.*, 2009; Chou *et al.*, 2007; Nayak *et al.*, 2010). Altogether these data strongly suggest that: (i) VSRs are *bona fide* virulence factors as they reduce viral pathogenicity when mutated or absent (Galiana-Arnoux *et al.*, 2006) but are sufficient to promote pathogenicity in an otherwise innocuous virus; (ii) RNAi reduces viral replication to below a threshold under which the virus is not pathogenic; this threshold is in turn reduced by VSRs and (iii) non-pathogenic viruses such as SINV have an intrinsic pathogenic potential that is revealed when the effects of RNAi are lessened. These conclusions are summarised in Fig. 3.

4.2 VIRAL DIVERSITY

The RNA-dependent RNA polymerases (RdRp) from RNA viruses lack proof-reading activity. Consequently, RNA viruses accumulate mutations at high frequency ($\sim 10^{-4}$/replication cycle, Sanjuan *et al.*, 2010). Therefore, the genome of viral samples cannot be averaged to its consensus sequence but should be considered as the sum of genomic variants present in the samples (Domingo *et al.*, 2006). The role of the accumulation of mutations as a possible RNAi escape mechanism has been indirectly observed in two different systems.

In vertebrate cells the transfection of synthetic siRNA directed against the viral consensus sequence inhibits viral replication while selecting genomic variants that present mutations in the sequence targeted by the siRNA (e.g. Boden *et al.*, 2003; Das *et al.*, 2004; Gitlin *et al.*, 2005); these mutations prevent perfect base-pairing between the siRNA and the viral RNA variants which then escape from degradation by the RNAi machinery. These results showed for the first time that mutations in a viral genome act as a countermeasure to RNAi-mediated degradation. However, these results were obtained *in vitro*, using organisms that naturally do not mount an antiviral RNAi response after infection.

In *Culex pipiens*, oral infection with West Nile virus triggers a natural RNAi response in the midgut of the infected mosquitoes (Brackney *et al.*, 2009). Using the sequence of vsiRNAs as a snapshot of the viral genomes being targeted by the RNAi machinery, the authors showed an increase in viral diversity through the course of the infection. This diversification in viral sequence was greater in regions of the West Nile virus genome that included vsiRNAs. In view of these results, the authors proposed that RNAi promotes viral population diversity. This tempting interpretation should be balanced by three main considerations: (i) the low coverage of vsiRNAs along the viral genome may result in a biased representation of genome diversity, especially in the less covered regions; (ii) the increased number of mutations observed in vsiRNAs may reflect an increase in viral diversity, but does not imply a role for RNAi in generating this diversity and (iii) no evidence showed that the vsiRNAs were actually loaded into siRISC, thus reflecting an active sequence-specific targeting of viral genomes.

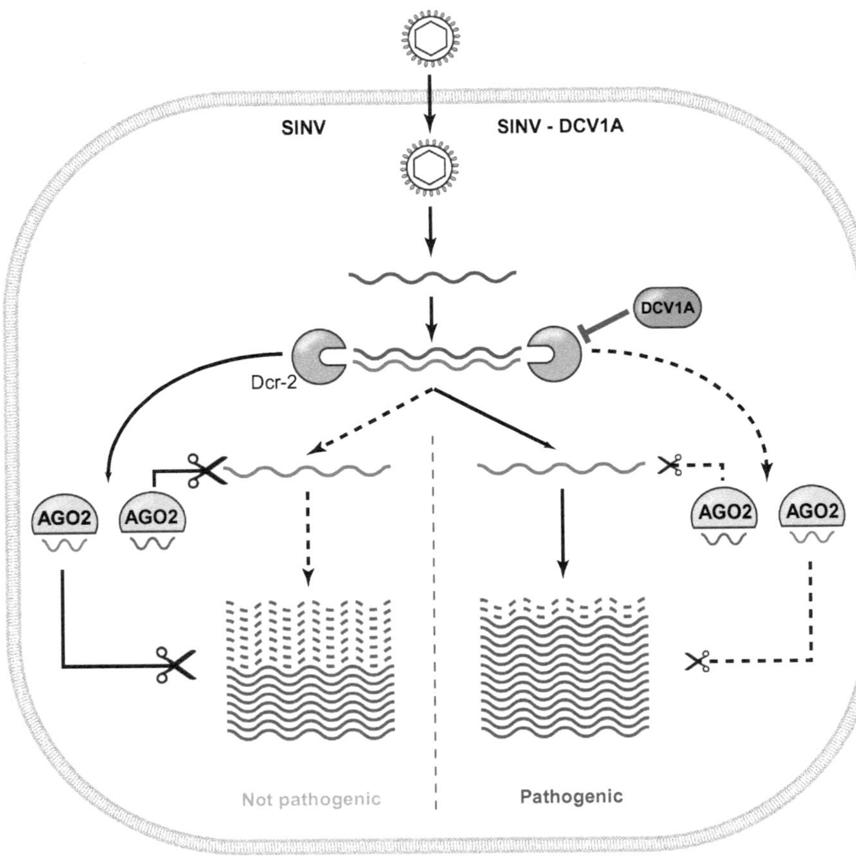

FIG. 3 Model illustrating the impact of VSR on RNAi, viral replication and outcome as exemplified by the SINV (left) and DCV1A-expressing SINV (right). Upon SINV infection (left side), the virus releases its positive strand genome (red) which is first translated and then used as a template for synthesis of the negative strand genome (green). In turn, the negative strand RNA serves as template for the synthesis of numerous positive strand RNAs that will be translated to increase the pool of viral proteins and encapsidated in newly synthesised viruses. Through this replication process, RNAi targets various molecular forms of the viral RNA. First, dsRNA replication intermediates (green and red) are cleaved by Dcr-2 and SINV-derived siRNAs are loaded into AGO2. These siRNAs guide the cleavage of viral RNAs of both polarity, limiting the accumulation of viral RNAs. Under these wild-type conditions, SINV is not pathogenic for insects. (Right side) The expression of the long dsRNA-specific binding protein DCV1A from the SINV genome protects the dsRNA intermediates of replication from degradation by Dcr-2. Consequently, fewer siRNAs are produced and loaded into AGO2, leading to a greater accumulation of viral RNAs. Under these conditions, DCV1A-expressing SINV is pathogenic for *Drosophila* contrary to the wild-type virus. (See Color Plate 3 at the back of the book.)

Nevertheless, data collected from both systems suggest a potential role of mutations in the viral genome as an escape mechanism from RNAi. Further experiments are necessary to directly assess the relationship between viral diversity and RNAi *in vivo*.

4.3 OTHER ESCAPE MECHANISMS

The relationship between RNAi and viral replication has only been studied at the molecular level regardless of the intracellular physiology of this interaction. Nevertheless, several aspects of viral replication may protect viral RNA from being degraded by the RNAi machinery. First, viral replication takes place at the membrane of virus-induced cellular compartments of various origins that may limit the accessibility of Dcr-2 to the dsRNA. Consequently, the antiviral activity of Dcr-2 would be limited to the subset of replication intermediates that escape from this protective niche. Interestingly, the site of replication may also influence the virus-RNAi relationship. The FHV replication complex is localised at the mitochondrial outer membrane in different hosts (Miller and Ahlquist, 2002; Miller *et al.*, 2001). In yeast, the retargeting of the FHV replisome to the endoplasmic reticulum results in increased RNA synthesis (Miller *et al.*, 2003) through an uncharacterised mechanism. This increased RNA synthesis could result from more efficient replication or better protection against degradation. These hypotheses remain to be investigated in light of antiviral RNAi in an organism, which contrary to *Saccharomyces cerevisiae*, possesses proficient RNAi machinery. Finally, translation and encapsidation may provide shelter for viral RNA by limiting their accessibility to AGO2.

5 Closing remarks

In the past decade, RNAi has been identified as a major mechanism of gene regulation with impacts on several cellular processes. In immunity, RNAi pathways play central roles in immune reactions both as regulators and as effectors. In insects, RNAi seems to be predominantly involved in antiviral defence through its direct antiviral activity; in vertebrates, RNAi has mostly been implicated in the regulation of immune responses. This last function for RNAi has not been addressed in insects but results obtained in other organisms suggest that such a role is conserved. Interestingly, while antiviral RNAi as defined in insects remains to be established in vertebrates, the intracellular recognition of dsRNA as a signature of viral infection is shared by both (Deddouche *et al.*, 2008). This is reminiscent of the antimicrobial response where the intracellular pathways are conserved between insects and vertebrates, while the upstream activation and the downstream effectors are different.

One of the most intriguing aspects of the antiviral RNAi in insects is its inability to eliminate the virus from the organism; in fact RNAi controls but never clears a viral infection. Whether controlling (functional efficiency) over eliminating (molecular efficiency) the virus is the ideal, rather than a compromise, for the insect is subject of debate. To address this question, let us consider the case of the SINV. Wild-type SINV is not pathogenic for insects in a siRNA pathway-dependent fashion. Therefore, the siRNA pathway is functionally efficient and controlling viral replication is sufficient to limit the deleterious effects of a viral infection. Nevertheless, the question of molecular efficiency remains unanswered. Several reasons may account for such an 'ineffectiveness'. First, it has been observed that genes encoding some components of the siRNA pathway display a more rapid evolutionary rate that non-immune genes presumably due to their targeting by viral suppressing factors. Therefore, the fast evolution of these genes prevents the siRNA pathway from becoming optimal in eradicating the virus while adapting to viral countermeasures. Second, it has been recently shown that the intracellular bacterium *Wolbachia* protects *Drosophila* from viral infection (Teixeira *et al.*, 2008) most likely in an RNAi-independent fashion. Moreover, it was suggested that the prevalence of *Wolbachia* infection in natural populations of *Drosophila* confers a selective advantage, presumably against infection by RNA viruses (Hedges *et al.*, 2008). This additional layer of defence may relax selection for a more molecularly efficient siRNA pathway by limiting viral replication upstream of or in parallel to it. Finally, some aspects of viral physiology such as encapsidation may render the clearance of the virus impossible. While most of the work has focused on the molecular players involved in host-virus interaction, further experiments are needed to evaluate the relationships between viral replication and RNAi in their physiological context.

Along the same lines, several aspects of RNAi biology remain to be addressed in insects including the following: (i) do miRNAs regulate immune responses, (ii) is there a role for RNAi in response to DNA viruses, (iii) what are the spatial and temporal relationships between viral replication and RNAi processing and (iv) can bacterial or fungal pathogens turn RNAi to their own benefit as observed for some viruses. Addressing these questions would provide a better understanding of the implication of small RNA pathways as surveillance, signalling and effector mechanisms for immunity.

Acknowledgements

We apologize to all of our colleagues whose work has not been cited due to space constraints. We thank the members of the Saleh lab for advice, suggestions and comments on the manuscript, Marco Vignuzzi and Ronald van Rij for critical reading of the manuscript, Bassam Berry and Christophe Antoniewski for sharing unpublished data and Elizabeth Jockusch for editorial support. This work was financially supported by the Agence Nationale de la Recherche (ANR-09-JCJC-0045-01) and the European Research Council (FP7/2007-2013 ERC 242703) to M.C.S.

Note added in proof

Due to the extensive time between writing and peer-review of the manuscript, its acceptance for publication (January 2011) and its final publication, some of the sections are not up to date. For example, we have recently demonstrated the production of virus-derived piRNAs following arbovirus infection in three mosquito cells lines: C6/36 and U4.4 from *A. albopictus*, and Aag2 from *A. aegypti* (Vodovar *et al.*, PlosOne, 2012, in press).

References

Adelman, Z. N., Blair, C. D., Carlson, J. O., Beaty, B. J. and Olson, K. E. (2001). Sindbis virus-induced silencing of dengue viruses in mosquitoes. *Insect Mol. Biol.* **10**, 265–273.

Adelman, Z. N., Sanchez-Vargas, I., Travanty, E. A., Carlson, J. O., Beaty, B. J., Blair, C. D. and Olson, K. E. (2002). RNA silencing of dengue virus type 2 replication in transformed C6/36 mosquito cells transcribing an inverted-repeat RNA derived from the virus genome. *J. Virol.* **76**, 12925–12933.

Agaisse, H., Petersen, U. M., Boutros, M., Mathey-Prevot, B. and Perrimon, N. (2003). Signaling role of hemocytes in Drosophila JAK/STAT-dependent response to septic injury. *Dev. Cell* **5**, 441–450.

Alder, M. N., Dames, S., Gaudet, J. and Mango, S. E. (2003). Gene silencing in Caenorhabditis elegans by transitive RNA interference. *RNA* **9**, 25–32.

Amdam, G. V., Simoes, Z. L., Guidugli, K. R., Norberg, K. and Omholt, S. W. (2003). Disruption of vitellogenin gene function in adult honeybees by intra-abdominal injection of double-stranded RNA. *BMC Biotechnol.* **3**, 1.

Anandalakshmi, R., Pruss, G. J., Ge, X., Marathe, R., Mallory, A. C., Smith, T. H. and Vance, V. B. (1998). A viral suppressor of gene silencing in plants. *Proc. Natl. Acad. Sci. USA* **95**, 13079–13084.

Araujo, R. N., Campos, I. T., Tanaka, A. S., Santos, A., Gontijo, N. F., Lehane, M. J. and Pereira, M. H. (2007). Brasiliensin: A novel intestinal thrombin inhibitor from *Triatoma brasiliensis* (Hemiptera: Reduviidae) with an important role in blood intake. *Int. J. Parasitol.* **37**, 1351–1358.

Araujo, R. N., Santos, A., Pinto, F. S., Gontijo, N. F., Lehane, M. J. and Pereira, M. H. (2006). RNA interference of the salivary gland nitrophorin 2 in the triatomine bug *Rhodnius prolixus* (Hemiptera: Reduviidae) by dsRNA ingestion or injection. *Insect Biochem. Mol. Biol.* **36**, 683–693.

Aravin, A. A., Lagos-Quintana, M., Yalcin, A., Zavolan, M., Marks, D., Snyder, B., Gaasterland, T., Meyer, J. and Tuschl, T. (2003). The small RNA profile during *Drosophila melanogaster* development. *Dev. Cell* **5**, 337–350.

Aravin, A. A., Hannon, G. J. and Brennecke, J. (2007a). The Piwi-piRNA pathway provides an adaptive defense in the transposon arms race. *Science* **318**, 761–764.

Aravin, A. A., Sachidanandam, R., Girard, A., Fejes-Toth, K. and Hannon, G. J. (2007b). Developmentally regulated piRNA clusters implicate MILI in transposon control. *Science* **316**, 744–747.

Arbouzova, N. I. and Zeidler, M. P. (2006). JAK/STAT signalling in Drosophila: insights into conserved regulatory and cellular functions. *Development* **133**, 2605–2616.

Attarzadeh-Yazdi, G., Fragkoudis, R., Chi, Y., Siu, R. W., Ulper, L., Barry, G., Rodriguez-Andres, J., Nash, A. A., Bouloy, M., Merits, A., Fazakerley, J. K. and Kohl, A. (2009). Cell-to-cell spread of the RNA interference response suppresses Semliki

Forest virus (SFV) infection of mosquito cell cultures and cannot be antagonized by SFV. *J. Virol.* **83**, 5735–5748.
Avadhanula, V., Weasner, B. P., Hardy, G. G., Kumar, J. P. and Hardy, R. W. (2009). A novel system for the launch of alphavirus RNA synthesis reveals a role for the Imd pathway in arthropod antiviral response. *PLoS Pathog.* **5**, e1000582.
Bartholomay, L. C., Fuchs, J. F., Cheng, L. L., Beck, E. T., Vizioli, J., Lowenberger, C. and Christensen, B. M. (2004). Reassessing the role of defensin in the innate immune response of the mosquito, *Aedes aegypti*. *Insect Mol. Biol.* **13**, 125–132.
Belyi, V. A., Levine, A. J. and Skalka, A. M. (2010). Unexpected inheritance: multiple integrations of ancient bornavirus and ebolavirus/marburgvirus sequences in vertebrate genomes. *PLoS Pathog.* **6**, e1001030.
Berry, B., Deddouche, S., Kirschner, D., Imler, J. L. and Antoniewski, C. (2009). Viral suppressors of RNA silencing hinder exogenous and endogenous small RNA pathways in Drosophila. *PLoS One* **4**, e5866.
Bettencourt, R., Terenius, O. and Faye, I. (2002). Hemolin gene silencing by ds-RNA injected into *Cecropia pupae* is lethal to next generation embryos. *Insect Mol. Biol.* **11**, 267–271.
Bier, E. (2005). *Drosophila*, the golden bug, emerges as a tool for human genetics. *Nat. Rev. Genet.* **6**, 9–23.
Billecocq, A., Vazeille-Falcoz, M., Rodhain, F. and Bouloy, M. (2000). Pathogen-specific resistance to Rift Valley fever virus infection is induced in mosquito cells by expression of the recombinant nucleoprotein but not NSs non-structural protein sequences. *J. Gen. Virol.* **81**, 2161–2166.
Blandin, S., Moita, L. F., Kocher, T., Wilm, M., Kafatos, F. C. and Levashina, E. A. (2002). Reverse genetics in the mosquito *Anopheles gambiae*: targeted disruption of the Defensin gene. *EMBO Rep.* **3**, 852–856.
Boden, D., Pusch, O., Lee, F., Tucker, L. and Ramratnam, B. (2003). Human immunodeficiency virus type 1 escape from RNA interference. *J. Virol.* **77**, 11531–11535.
Boman, H. G., Nilsson, I. and Rasmuson, B. (1972). Inducible antibacterial defence system in Drosophila. *Nature* **237**, 232–235.
Brackney, D. E., Beane, J. E. and Ebel, G. D. (2009). RNAi targeting of West Nile virus in mosquito midguts promotes virus diversification. *PLoS Pathog.* **5**, e1000502.
Brackney, D. E., Scott, J. C., Sagawa, F., Woodward, J. E., Miller, N. A., Schilkey, F. D., Mudge, J., Wilusz, J., Olson, K. E., Blair, C. D. and Ebel, G. D. (2010). C6/36 Aedes albopictus cells have a dysfunctional antiviral RNA interference response. *PLoS Negl. Trop. Dis.* **4**, e856.
Brennecke, J., Aravin, A. A., Stark, A., Dus, M., Kellis, M., Sachidanandam, R. and Hannon, G. J. (2007). Discrete small RNA-generating loci as master regulators of transposon activity in *Drosophila*. *Cell* **128**, 1089–1103.
Brigneti, G., Voinnet, O., Li, W. X., Ji, L. H., Ding, S. W. and Baulcombe, D. C. (1998). Viral pathogenicity determinants are suppressors of transgene silencing in *Nicotiana benthamiana*. *EMBO J.* **17**, 6739–6746.
Bucher, G., Scholten, J. and Klingler, M. (2002). Parental RNAi in Tribolium (Coleoptera). *Curr. Biol.* **12**, R85–86.
Campbell, C. L., Keene, K. M., Brackney, D. E., Olson, K. E., Blair, C. D., Wilusz, J. and Foy, B. D. (2008). Aedes aegypti uses RNA interference in defense against Sindbis virus infection. *BMC Microbiol.* **8**, 47.
Caudy, A. A., Myers, M., Hannon, G. J. and Hammond, S. M. (2002). Fragile X-related protein and VIG associate with the RNA interference machinery. *Genes Dev.* **16**, 2491–2496.
Chen, J., Zhang, D., Yao, Q., Zhang, J., Dong, X., Tian, H. and Zhang, W. (2010). Feeding-based RNA interference of a trehalose phosphate synthase gene in the brown planthopper, *Nilaparvata lugens*. *Insect Mol. Biol.* **19**, 777–786.

Chen, S. L., Dai, S. M., Lu, K. H. and Chang, C. (2008). Female-specific doublesex dsRNA interrupts yolk protein gene expression and reproductive ability in oriental fruit fly, Bactrocera dorsalis (Hendel). *Insect Biochem. Mol. Biol.* **38**, 155–165.

Chotkowski, H. L., Ciota, A. T., Jia, Y., Puig-Basagoiti, F., Kramer, L. D., Shi, P. Y. and Glaser, R. L. (2008). West Nile virus infection of Drosophila melanogaster induces a protective RNAi response. *Virology* **377**, 197–206.

Chou, Y. T., Tam, B., Linay, F. and Lai, E. C. (2007). Transgenic inhibitors of RNA interference in Drosophila. *Fly (Austin)* **1**, 311–316.

Chung, W. J., Okamura, K., Martin, R. and Lai, E. C. (2008). Endogenous RNA interference provides a somatic defense against Drosophila transposons. *Curr. Biol.* **18**, 795–802.

Cirimotich, C. M., Scott, J. C., Phillips, A. T., Geiss, B. J. and Olson, K. E. (2009). Suppression of RNA interference increases alphavirus replication and virus-associated mortality in Aedes aegypti mosquitoes. *BMC Microbiol.* **9**, 49.

Costa, A., Jan, E., Sarnow, P. and Schneider, D. (2009). The Imd pathway is involved in antiviral immune responses in Drosophila. *PLoS One* **4**, e7436.

Cox-Foster, D. L., Conlan, S., Holmes, E. C., Palacios, G., Evans, J. D., Moran, N. A., Quan, P. L., Briese, T., Hornig, M., Geiser, D. M., Martinson, V., vanEngelsdorp, D., *et al.* (2007). A metagenomic survey of microbes in honey bee colony collapse disorder. *Science* **318**, 283–287.

Crochu, S., Cook, S., Attoui, H., Charrel, R. N., De Chesse, R., Belhouchet, M., Lemasson, J. J., de Micco, P. and de Lamballerie, X. (2004). Sequences of flavivirus-related RNA viruses persist in DNA form integrated in the genome of Aedes spp. mosquitoes. *J. Gen. Virol.* **85**, 1971–1980.

Cullen, B. R. (2010). Five questions about viruses and microRNAs. *PLoS Pathog.* **6**, e1000787.

Das, A. T., Brummelkamp, T. R., Westerhout, E. M., Vink, M., Madiredjo, M., Bernards, R. and Berkhout, B. (2004). Human immunodeficiency virus type 1 escapes from RNA interference-mediated inhibition. *J. Virol.* **78**, 2601–2605.

Deddouche, S., Matt, N., Budd, A., Mueller, S., Kemp, C., Galiana-Arnoux, D., Dostert, C., Antoniewski, C., Hoffmann, J. A. and Imler, J. L. (2008). The DExD/H-box helicase Dicer-2 mediates the induction of antiviral activity in drosophila. *Nat. Immunol.* **9**, 1425–1432.

Denli, A. M., Tops, B. B., Plasterk, R. H., Ketting, R. F. and Hannon, G. J. (2004). Processing of primary microRNAs by the Microprocessor complex. *Nature* **432**, 231–235.

Domingo, E., Martin, V., Perales, C., Grande-Perez, A., Garcia-Arriaza, J. and Arias, A. (2006). Viruses as quasispecies: biological implications. *Curr. Top. Microbiol. Immunol.* **299**, 51–82.

Dong, Y. and Friedrich, M. (2005). Nymphal RNAi: systemic RNAi mediated gene knockdown in juvenile grasshopper. *BMC Biotechnol.* **5**, 25.

Dostert, C., Jouanguy, E., Irving, P., Troxler, L., Galiana-Arnoux, D., Hetru, C., Hoffmann, J. A. and Imler, J. L. (2005). The Jak-STAT signaling pathway is required but not sufficient for the antiviral response of drosophila. *Nat. Immunol.* **6**, 946–953.

Dzitoyeva, S., Dimitrijevic, N. and Manev, H. (2001). Intra-abdominal injection of double-stranded RNA into anesthetized adult Drosophila triggers RNA interference in the central nervous system. *Mol. Psychiatry* **6**, 665–670.

Eleftherianos, I., Millichap, P. J., ffrench-Constant, R. H. and Reynolds, S. E. (2006). RNAi suppression of recognition protein mediated immune responses in the tobacco hornworm Manduca sexta causes increased susceptibility to the insect pathogen Photorhabdus. *Dev. Comp. Immunol.* **30**, 1099–1107.

Eulalio, A., Huntzinger, E. and Izaurralde, E. (2008). Getting to the root of miRNA-mediated gene silencing. *Cell* **132**, 9–14.

Fagegaltier, D., Bouge, A. L., Berry, B., Poisot, E., Sismeiro, O., Coppee, J. Y., Theodore, L., Voinnet, O. and Antoniewski, C. (2009). The endogenous siRNA pathway is involved in heterochromatin formation in Drosophila. *Proc. Natl. Acad. Sci. USA* **106**, 21258–21263.

Feinberg, E. H. and Hunter, C. P. (2003). Transport of dsRNA into cells by the transmembrane protein SID-1. *Science* **301**, 1545–1547.

Fire, A., Xu, S., Montgomery, M. K., Kostas, S. A., Driver, S. E. and Mello, C. C. (1998). Potent and specific genetic interference by double-stranded RNA in Caenorhabditis elegans. *Nature* **391**, 806–811.

Flynt, A., Liu, N., Martin, R. and Lai, E. C. (2009). Dicing of viral replication intermediates during silencing of latent Drosophila viruses. *Proc. Natl. Acad. Sci. USA* **106**, 5270–5275.

Frank, A. C. and Wolfe, K. H. (2009). Evolutionary capture of viral and plasmid DNA by yeast nuclear chromosomes. *Eukaryot. Cell* **8**, 1521–1531.

Franz, A. W., Sanchez-Vargas, I., Adelman, Z. N., Blair, C. D., Beaty, B. J., James, A. A. and Olson, K. E. (2006). Engineering RNA interference-based resistance to dengue virus type 2 in genetically modified Aedes aegypti. *Proc. Natl. Acad. Sci. USA* **103**, 4198–4203.

Franz, A. W., Sanchez-Vargas, I., Piper, J., Smith, M. R., Khoo, C. C., James, A. A. and Olson, K. E. (2009). Stability and loss of a virus resistance phenotype over time in transgenic mosquitoes harbouring an antiviral effector gene. *Insect Mol. Biol.* **18**, 661–672.

Gaines, P. J., Olson, K. E., Higgs, S., Powers, A. M., Beaty, B. J. and Blair, C. D. (1996). Pathogen-derived resistance to dengue type 2 virus in mosquito cells by expression of the premembrane coding region of the viral genome. *J. Virol.* **70**, 2132–2137.

Galiana-Arnoux, D., Dostert, C., Schneemann, A., Hoffmann, J. A. and Imler, J. L. (2006). Essential function in vivo for Dicer-2 in host defense against RNA viruses in drosophila. *Nat. Immunol.* **7**, 590–597.

Geuking, M. B., Weber, J., Dewannieux, M., Gorelik, E., Heidmann, T., Hengartner, H., Zinkernagel, R. M. and Hangartner, L. (2009). Recombination of retrotransposon and exogenous RNA virus results in nonretroviral cDNA integration. *Science* **323**, 393–396.

Ghanim, M., Kontsedalov, S. and Czosnek, H. (2007). Tissue-specific gene silencing by RNA interference in the whitefly Bemisia tabaci (Gennadius). *Insect Biochem. Mol. Biol.* **37**, 732–738.

Ghildiyal, M., Seitz, H., Horwich, M. D., Li, C., Du, T., Lee, S., Xu, J., Kittler, E. L., Zapp, M. L., Weng, Z. and Zamore, P. D. (2008). Endogenous siRNAs derived from transposons and mRNAs in Drosophila somatic cells. *Science* **320**, 1077–1081.

Ghildiyal, M., Xu, J., Seitz, H., Weng, Z. and Zamore, P. D. (2010). Sorting of Drosophila small silencing RNAs partitions microRNA* strands into the RNA interference pathway. *RNA* **16**, 43–56.

Gitlin, L., Stone, J. K. and Andino, R. (2005). Poliovirus escape from RNA interference: short interfering RNA-target recognition and implications for therapeutic approaches. *J. Virol.* **79**, 1027–1035.

Gottar, M., Gobert, V., Matskevich, A. A., Reichhart, J. M., Wang, C., Butt, T. M., Belvin, M., Hoffmann, J. A. and Ferrandon, D. (2006). Dual detection of fungal infections in Drosophila via recognition of glucans and sensing of virulence factors. *Cell* **127**, 1425–1437.

Hammond, S. M., Boettcher, S., Caudy, A. A., Kobayashi, R. and Hannon, G. J. (2001). Argonaute2, a link between genetic and biochemical analyses of RNAi. *Science* **293**, 1146–1150.

Han, J., Lee, Y., Yeom, K. H., Kim, Y. K., Jin, H. and Kim, V. N. (2004). The Drosha-DGCR8 complex in primary microRNA processing. *Genes Dev.* **18**, 3016–3027.

He, Z. B., Cao, Y. Q., Yin, Y. P., Wang, Z. K., Chen, B., Peng, G. X. and Xia, Y. X. (2006). Role of hunchback in segment patterning of Locusta migratoria manilensis revealed by parental RNAi. *Dev. Growth Differ.* **48**, 439–445.

Hedges, L. M. and Johnson, K. N. (2008). Induction of host defence responses by Drosophila C virus. *J. Gen. Virol.* **89**, 1497–1501.

Hedges, L. M., Brownlie, J. C., O'Neill, S. L. and Johnson, K. N. (2008). Wolbachia and virus protection in insects. *Science* **322**, 702.

Horie, M., Honda, T., Suzuki, Y., Kobayashi, Y., Daito, T., Oshida, T., Ikuta, K., Jern, P., Gojobori, T., Coffin, J. M. and Tomonaga, K. (2010). Endogenous non-retroviral RNA virus elements in mammalian genomes. *Nature* **463**, 84–87.

Houwing, S., Kamminga, L. M., Berezikov, E., Cronembold, D., Girard, A., van den Elst, H., Filippov, D. V., Blaser, H., Raz, E., Moens, C. B., Plasterk, R. H., Hannon, G. J., *et al.* (2007). A role for Piwi and piRNAs in germ cell maintenance and transposon silencing in Zebrafish. *Cell* **129**, 69–82.

Infanger, L. C., Rocheleau, T. A., Bartholomay, L. C., Johnson, J. K., Fuchs, J., Higgs, S., Chen, C. C. and Christensen, B. M. (2004). The role of phenylalanine hydroxylase in melanotic encapsulation of filarial worms in two species of mosquitoes. *Insect Biochem. Mol. Biol.* **34**, 1329–1338.

Iwasaki, S., Kawamata, T. and Tomari, Y. (2009). Drosophila argonaute1 and argonaute2 employ distinct mechanisms for translational repression. *Mol. Cell* **34**, 58–67.

Janeway, C. A. Jr. (1989). Approaching the asymptote? Evolution and revolution in immunology. *Cold Spring Harb. Symp. Quant. Biol.* **54**(Pt 1), 1–13.

Janeway, C. A. Jr. (1992). The immune system evolved to discriminate infectious nonself from noninfectious self. *Immunol. Today* **13**, 11–16.

John, B., Enright, A. J., Aravin, A., Tuschl, T., Sander, C. and Marks, D. S. (2004). Human MicroRNA targets. *PLoS Biol.* **2**, e363.

Jaubert-Possamai, S., Le Trionnaire, G., Bonhomme, J., Christophides, G. K., Rispe, C. and Tagu, D. (2007). Gene knockdown by RNAi in the pea aphid Acyrthosiphon pisum. *BMC Biotechnol.* **7**, 63.

Kasschau, K. D. and Carrington, J. C. (1998). A counterdefensive strategy of plant viruses: suppression of posttranscriptional gene silencing. *Cell* **95**, 461–470.

Keene, K. M., Foy, B. D., Sanchez-Vargas, I., Beaty, B. J., Blair, C. D. and Olson, K. E. (2004). RNA interference acts as a natural antiviral response to O'nyong-nyong virus (Alphavirus; Togaviridae) infection of Anopheles gambiae. *Proc. Natl. Acad. Sci. USA* **101**, 17240–17245.

Khoo, C. C., Piper, J., Sanchez-Vargas, I., Olson, K. E. and Franz, A. W. (2010). The RNA interference pathway affects midgut infection- and escape barriers for Sindbis virus in Aedes aegypti. *BMC Microbiol.* **10**, 130.

Lavie, P., Fresnaye, J. and Vago, C. (1965). L'action de la virose à noyaux denses sur les larves de Galleria mellonella dans les ruches. *Ann. Abeille* **8**, 321–323.

Lazzaro, B. P. (2008). Natural selection on the Drosophila antimicrobial immune system. *Curr. Opin. Microbiol.* **11**, 284–289.

Lee, Y., Ahn, C., Han, J., Choi, H., Kim, J., Yim, J., Lee, J., Provost, P., Radmark, O., Kim, S. and Kim, V. N. (2003). The nuclear RNase III Drosha initiates microRNA processing. *Nature* **425**, 415–419.

Lee, Y. S., Nakahara, K., Pham, J. W., Kim, K., He, Z., Sontheimer, E. J. and Carthew, R. W. (2004). Distinct roles for Drosophila Dicer-1 and Dicer-2 in the siRNA/miRNA silencing pathways. *Cell* **117**, 69–81.

Lehane, M. J., Gibson, W. and Lehane, S. M. (2008). Differential expression of fat body genes in Glossina morsitans morsitans following infection with Trypanosoma brucei brucei. *Int. J. Parasitol.* **38**, 93–101.

Lemaitre, B. (2004). The road to Toll. *Nat. Rev. Immunol.* **4**, 521–527.

Lemaitre, B. and Hoffmann, J. (2007). The host defense of Drosophila melanogaster. *Annu. Rev. Immunol.* **25**, 697–743.

Lemaitre, B., Kromer-Metzger, E., Michaut, L., Nicolas, E., Meister, M., Georgel, P., Reichhart, J. M. and Hoffmann, J. A. (1995a). A recessive mutation, immune deficiency (imd), defines two distinct control pathways in the Drosophila host defense. *Proc. Natl. Acad. Sci. USA* **92**, 9465–9469.

Lemaitre, B., Meister, M., Govind, S., Georgel, P., Steward, R., Reichhart, J. M. and Hoffmann, J. A. (1995b). Functional analysis and regulation of nuclear import of dorsal during the immune response in Drosophila. *EMBO J.* **14**, 536–545.

Lemaitre, B., Nicolas, E., Michaut, L., Reichhart, J. M. and Hoffmann, J. A. (1996). The dorsoventral regulatory gene cassette spatzle/Toll/cactus controls the potent antifungal response in Drosophila adults. *Cell* **86**, 973–983.

Lemaitre, B., Reichhart, J. M. and Hoffmann, J. A. (1997). Drosophila host defense: differential induction of antimicrobial peptide genes after infection by various classes of microorganisms. *Proc. Natl. Acad. Sci. USA* **94**, 14614–14619.

Leulier, F., Parquet, C., Pili-Floury, S., Ryu, J. H., Caroff, M., Lee, W. J., Mengin-Lecreulx, D. and Lemaitre, B. (2003). The Drosophila immune system detects bacteria through specific peptidoglycan recognition. *Nat. Immunol.* **4**, 478–484.

Li, F. and Ding, S. W. (2006). Virus counterdefense: diverse strategies for evading the RNA-silencing immunity. *Annu. Rev. Microbiol.* **60**, 503–531.

Li, H. W., Lucy, A. P., Guo, H. S., Li, W. X., Ji, L. H., Wong, S. M. and Ding, S. W. (1999). Strong host resistance targeted against a viral suppressor of the plant gene silencing defence mechanism. *EMBO J.* **18**, 2683–2691.

Li, H., Li, W. X. and Ding, S. W. (2002). Induction and suppression of RNA silencing by an animal virus. *Science* **296**, 1319–1321.

Li, S. C., Shiau, C. K. and Lin, W. C. (2008). Vir-Mir db: prediction of viral microRNA candidate hairpins. *Nucleic Acids Res.* **36**, D184–D189.

Li, Y., Chan, E. Y., Li, J., Ni, C., Peng, X., Rosenzweig, E., Tumpey, T. M. and Katze, M. G. (2010). MicroRNA expression and virulence in pandemic influenza virus-infected mice. *J. Virol.* **84**, 3023–3032.

Lipardi, C. and Paterson, B. M. (2009). Identification of an RNA-dependent RNA polymerase in Drosophila involved in RNAi and transposon suppression. *Proc. Natl. Acad. Sci. USA* **106**, 15645–15650.

Liu, S., Ding, Z., Zhang, C., Yang, B. and Liu, Z. (2010). Gene knockdown by intro-thoracic injection of double-stranded RNA in the brown planthopper, Nilaparvata lugens. *Insect Biochem. Mol. Biol.* **40**, 666–671.

Liu, Q., Rand, T. A., Kalidas, S., Du, F., Kim, H. E., Smith, D. P. and Wang, X. (2003). R2D2, a bridge between the initiation and effector steps of the Drosophila RNAi pathway. *Science* **301**, 1921–1925.

Liu, J., Carmell, M. A., Rivas, F. V., Marsden, C. G., Thomson, J. M., Song, J. J., Hammond, S. M., Joshua-Tor, L. and Hannon, G. J. (2004). Argonaute2 is the catalytic engine of mammalian RNAi. *Science* **305**, 1437–1441.

Liu, X., Jiang, F., Kalidas, S., Smith, D. and Liu, Q. (2006). Dicer-2 and R2D2 coordinately bind siRNA to promote assembly of the siRISC complexes. *RNA* **12**, 1514–1520.

Liu, Y., Ye, X., Jiang, F., Liang, C., Chen, D., Peng, J., Kinch, L. N., Grishin, N. V. and Liu, Q. (2009). C3PO, an endoribonuclease that promotes RNAi by facilitating RISC activation. *Science* **325**, 750–753.

Lu, R., Folimonov, A., Shintaku, M., Li, W. X., Falk, B. W., Dawson, W. O. and Ding, S. W. (2004). Three distinct suppressors of RNA silencing encoded by a 20-kb viral RNA genome. *Proc. Natl. Acad. Sci. USA* **101**, 15742–15747.

Lu, R., Maduro, M., Li, F., Li, H. W., Broitman-Maduro, G., Li, W. X. and Ding, S. W. (2005). Animal virus replication and RNAi-mediated antiviral silencing in Caenorhabditis elegans. *Nature* **436**, 1040–1043.

Lungchukiet, P., Donly, B. C., Zhang, J., Tobe, S. S. and Bendena, W. G. (2008). Molecular cloning and characterization of an allatostatin-like receptor in the cockroach Diploptera punctata. *Peptides* **29**, 276–285.

Lynch, J. A., Brent, A. E., Leaf, D. S., Pultz, M. A. and Desplan, C. (2006). Localized maternal orthodenticle patterns anterior and posterior in the long germ wasp Nasonia. *Nature* **439**, 728–732.

Makeyev, E. V. and Bamford, D. H. (2002). Cellular RNA-dependent RNA polymerase involved in posttranscriptional gene silencing has two distinct activity modes. *Mol. Cell* **10**, 1417–1427.

Maori, E., Tanne, E. and Sela, I. (2007). Reciprocal sequence exchange between non-retro viruses and hosts leading to the appearance of new host phenotypes. *Virology* **362**, 342–349.

Maori, E., Paldi, N., Shafir, S., Kalev, H., Tsur, E., Glick, E. and Sela, I. (2009). IAPV, a bee-affecting virus associated with Colony Collapse Disorder can be silenced by dsRNA ingestion. *Insect Mol. Biol.* **18**, 55–60.

Marques, J. T., Kim, K., Wu, P. H., Alleyne, T. M., Jafari, N. and Carthew, R. W. (2010). Loqs and R2D2 act sequentially in the siRNA pathway in Drosophila. *Nat. Struct. Mol. Biol.* **17**, 24–30.

Martin, D., Maestro, O., Cruz, J., Mane-Padros, D. and Belles, X. (2006). RNAi studies reveal a conserved role for RXR in molting in the cockroach Blattella germanica. *J. Insect Physiol.* **52**, 410–416.

Matranga, C., Tomari, Y., Shin, C., Bartel, D. P. and Zamore, P. D. (2005). Passenger-strand cleavage facilitates assembly of siRNA into Ago2-containing RNAi enzyme complexes. *Cell* **123**, 607–620.

Matzinger, P. (1994). Tolerance, danger, and the extended family. *Annu. Rev. Immunol.* **12**, 991–1045.

Matzinger, P. (2007). Friendly and dangerous signals: is the tissue in control? *Nat. Immunol.* **8**, 11–13.

Mellor, P. S. (2000). Replication of arboviruses in insect vectors. *J. Comp. Pathol.* **123**, 231–247.

Miller, D. J. and Ahlquist, P. (2002). Flock house virus RNA polymerase is a transmembrane protein with amino-terminal sequences sufficient for mitochondrial localization and membrane insertion. *J. Virol.* **76**, 9856–9867.

Miller, D. J., Schwartz, M. D. and Ahlquist, P. (2001). Flock house virus RNA replicates on outer mitochondrial membranes in Drosophila cells. *J. Virol.* **75**, 11664–11676.

Miller, D. J., Schwartz, M. D., Dye, B. T. and Ahlquist, P. (2003). Engineered retargeting of viral RNA replication complexes to an alternative intracellular membrane. *J. Virol.* **77**, 12193–12202.

Miyoshi, T., Takeuchi, A., Siomi, H. and Siomi, M. C. (2010). A direct role for Hsp90 in pre-RISC formation in Drosophila. *Nat. Struct. Mol. Biol.* **17**, 1024–1026.

Morgan, T. H. (1910). Sex limited inheritance in Drosophila. *Science* **32**, 120–122.

Mueller, S., Gausson, V., Vodovar, N., Deddouche, S., Troxler, L., Perot, J., Pfeffer, S., Hoffmann, J. A., Saleh, M. C. and Imler, J. L. (2010). RNAi-mediated immunity provides strong protection against the negative-strand RNA vesicular stomatitis virus in Drosophila. *Proc. Natl. Acad. Sci. USA* **107**, 19390–19395.

Meyering-Vos, M., Merz, S., Sertkol, M. and Hoffmann, K. H. (2006). Functional analysis of the allatostatin-A type gene in the cricket Gryllus bimaculatus and the armyworm Spodoptera frugiperda. *Insect Biochem. Mol. Biol.* **36**, 492–504.

Nayak, A., Berry, B., Tassetto, M., Kunitomi, M., Acevedo, A., Deng, C., Krutchinsky, A., Gross, J., Antoniewski, C. and Andino, R. (2010). Cricket paralysis virus antagonizes Argonaute 2 to modulate antiviral defense in Drosophila. *Nat. Struct. Mol. Biol.* **17**, 547–554.

Niu, B. L., Shen, W. F., Liu, Y., Weng, H. B., He, L. H., Mu, J. J., Wu, Z. L., Jiang, P., Tao, Y. Z. and Meng, Z. Q. (2008). Cloning and RNAi-mediated functional characterization of MaLac2 of the pine sawyer, Monochamus alternatus. *Insect Mol. Biol.* **17**, 303–312.

Obbard, D. J., Jiggins, F. M., Halligan, D. L. and Little, T. J. (2006). Natural selection drives extremely rapid evolution in antiviral RNAi genes. *Curr. Biol.* **16**, 580–585.

Olivieri, D., Sykora, M. M., Sachidanandam, R., Mechtler, K. and Brennecke, J. (2010). An in vivo RNAi assay identifies major genetic and cellular requirements for primary piRNA biogenesis in Drosophila. *EMBO J.* **29**, 3301–3317.

Olson, K. E., Higgs, S., Gaines, P. J., Powers, A. M., Davis, B. S., Kamrud, K. I., Carlson, J. O., Blair, C. D. and Beaty, B. J. (1996). Genetically engineered resistance to dengue-2 virus transmission in mosquitoes. *Science* **272**, 884–886.

Ohde, T., Masumoto, M., Morita-Miwa, M., Matsuura, H., Yoshioka, H., Yaginuma, T. and Niimi, T. (2009). Vestigial and scalloped in the ladybird beetle: a conserved function in wing development and a novel function in pupal ecdysis. *Insect Mol. Biol.* **18**, 571–581.

Pak, J. and Fire, A. (2007). Distinct populations of primary and secondary effectors during RNAi in C. elegans. *Science* **315**, 241–244.

Parameswaran, P., Sklan, E., Wilkins, C., Burgon, T., Samuel, M. A., Lu, R., Ansel, K. M., Heissmeyer, V., Einav, S., Jackson, W., Doukas, T., Paranjape, S., *et al.* (2010). Six RNA viruses and forty-one hosts: viral small RNAs and modulation of small RNA repertoires in vertebrate and invertebrate systems. *PLoS Pathog.* **6**, e1000764.

Patel, A., Fondrk, M. K., Kaftanoglu, O., Emore, C., Hunt, G., Frederick, K. and Amdam, G. V. (2007). The making of a queen: TOR pathway is a key player in diphenic caste development. *PLoS One* **2**, e509.

Pfeffer, S., Sewer, A., Lagos-Quintana, M., Sheridan, R., Sander, C., Grasser, F. A., van Dyk, L. F., Ho, C. K., Shuman, S., Chien, M., Russo, J. J., Ju, J., *et al.* (2005). Identification of microRNAs of the herpesvirus family. *Nat. Methods* **2**, 269–276.

Rand, T. A., Petersen, S., Du, F. and Wang, X. (2005). Argonaute2 cleaves the anti-guide strand of siRNA during RISC activation. *Cell* **123**, 621–629.

Ratcliff, F., Harrison, B. D. and Baulcombe, D. C. (1997). A similarity between viral defense and gene silencing in plants. *Science* **276**, 1558–1560.

Rajagopal, R., Sivakumar, S., Agrawal, N., Malhotra, P. and Bhatnagar, R. K. (2002). Silencing of midgut aminopeptidase N of Spodoptera litura by double-stranded RNA establishes its role as Bacillus thuringiensis toxin receptor. *J. Biol. Chem.* **277**, 46849–46851.

Rock, F. L., Hardiman, G., Timans, J. C., Kastelein, R. A. and Bazan, J. F. (1998). A family of human receptors structurally related to Drosophila Toll. *Proc. Natl. Acad. Sci. USA* **95**, 588–593.

Roignant, J. Y., Carre, C., Mugat, B., Szymczak, D., Lepesant, J. A. and Antoniewski, C. (2003). Absence of transitive and systemic pathways allows cell-specific and isoform-specific RNAi in Drosophila. *RNA* **9**, 299–308.

Rouha, H., Thurner, C. and Mandl, C. W. (2010). Functional microRNA generated from a cytoplasmic RNA virus. *Nucleic Acids Res.* **38**, 8328–8337.

Sant'Anna, M. R., Alexander, B., Bates, P. A. and Dillon, R. J. (2008). Gene silencing in phlebotomine sand flies: Xanthine dehydrogenase knock down by dsRNA microinjections. *Insect Biochem. Mol. Biol.* **38**, 652–660.

Sabin, L. R., Zhou, R., Gruber, J. J., Lukinova, N., Bambina, S., Berman, A., Lau, C. K., Thompson, C. B. and Cherry, S. (2009). Ars2 regulates both miRNA- and siRNA-dependent silencing and suppresses RNA virus infection in Drosophila. *Cell* **138**, 340–351.

Saleh, M. C., van Rij, R. P., Hekele, A., Gillis, A., Foley, E., O'Farrell, P. H. and Andino, R. (2006). The endocytic pathway mediates cell entry of dsRNA to induce RNAi silencing. *Nat. Cell Biol.* **8**, 793–802.

Saleh, M. C., Tassetto, M., van Rij, R. P., Goic, B., Gausson, V., Berry, B., Jacquier, C., Antoniewski, C. and Andino, R. (2009). Antiviral immunity in Drosophila requires systemic RNA interference spread. *Nature* **458**, 346–350.

Sanchez-Vargas, I., Scott, J. C., Poole-Smith, B. K., Franz, A. W., Barbosa-Solomieu, V., Wilusz, J., Olson, K. E. and Blair, C. D. (2009). Dengue virus type 2 infections of Aedes aegypti are modulated by the mosquito's RNA interference pathway. *PLoS Pathog.* **5**, e1000299.

Sanjuan, R., Nebot, M. R., Chirico, N., Mansky, L. M. and Belshaw, R. (2010). Viral mutation rates. *J. Virol.* **84**, 9733–9748.

Schott, D. H., Cureton, D. K., Whelan, S. P. and Hunter, C. P. (2005). An antiviral role for the RNA interference machinery in Caenorhabditis elegans. *Proc. Natl. Acad. Sci. USA* **102**, 18420–18424.

Schwarz, D. S., Hutvagner, G., Du, T., Xu, Z., Aronin, N. and Zamore, P. D. (2003). Asymmetry in the assembly of the RNAi enzyme complex. *Cell* **115**, 199–208.

Scott, J. C., Brackney, D. E., Campbell, C. L., Bondu-Hawkins, V., Hjelle, B., Ebel, G. D., Olson, K. E. and Blair, C. D. (2010). Comparison of dengue virus type 2-specific small RNAs from RNA interference-competent and -incompetent mosquito cells. *PLoS Negl. Trop. Dis.* **4**, e848.

Sessions, O. M., Barrows, N. J., Souza-Neto, J. A., Robinson, T. J., Hershey, C. L., Rodgers, M. A., Ramirez, J. L., Dimopoulos, G., Yang, P. L., Pearson, J. L. and Garcia-Blanco, M. A. (2009). Discovery of insect and human dengue virus host factors. *Nature* **458**, 1047–1050.

Sijen, T., Fleenor, J., Simmer, F., Thijssen, K. L., Parrish, S., Timmons, L., Plasterk, R. H. and Fire, A. (2001). On the role of RNA amplification in dsRNA-triggered gene silencing. *Cell* **107**, 465–476.

Silhavy, D., Molnar, A., Lucioli, A., Szittya, G., Hornyik, C., Tavazza, M. and Burgyan, J. (2002). A viral protein suppresses RNA silencing and binds silencing-generated, 21- to 25-nucleotide double-stranded RNAs. *EMBO J.* **21**, 3070–3080.

Singh, J., Singh, C. P., Bhavani, A. and Nagaraju, J. (2010). Discovering microRNAs from Bombyx mori nucleopolyhedrosis virus. *Virology* **407**, 120–128.

Sim, C. and Denlinger, D. L. (2008). Insulin signaling and FOXO regulate the overwintering diapause of the mosquito Culex pipiens. *Proc. Natl. Acad. Sci. USA* **105**, 6777–6781.

Smardon, A., Spoerke, J. M., Stacey, S. C., Klein, M. E., Mackin, N. and Maine, E. M. (2000). EGO-1 is related to RNA-directed RNA polymerase and functions in germ-line development and RNA interference in C. elegans. *Curr. Biol.* **10**, 169–178.

Souza-Neto, J. A., Sim, S. and Dimopoulos, G. (2009). An evolutionary conserved function of the JAK-STAT pathway in anti-dengue defense. *Proc. Natl. Acad. Sci. USA* **106**, 17841–17846.

Specchia, V., Benna, C., Mazzotta, G. M., Piccin, A., Zordan, M. A., Costa, R. and Bozzetti, M. P. (2008). aubergine gene overexpression in somatic tissues of aubergine (sting) mutants interferes with the RNAi pathway of a yellow hairpin dsRNA in Drosophila melanogaster. *Genetics* **178**, 1271–1282.

Sivakumar, S., Rajagopal, R., Venkatesh, G. R., Srivastava, A. and Bhatnagar, R. K. (2007). Knockdown of aminopeptidase-N from Helicoverpa armigera larvae and in transfected Sf21 cells by RNA interference reveals its functional interaction with Bacillus thuringiensis insecticidal protein Cry1Ac. *J. Biol. Chem.* **282**, 7312–7319.

Tam, O. H., Aravin, A. A., Stein, P., Girard, A., Murchison, E. P., Cheloufi, S., Hodges, E., Anger, M., Sachidanandam, R., Schultz, R. M. and Hannon, G. J. (2008). Pseudogene-derived small interfering RNAs regulate gene expression in mouse oocytes. *Nature* **453**, 534–538.

Tabunoki, H., Higurashi, S., Ninagi, O., Fujii, H., Banno, Y., Nozaki, M., Kitajima, M., Miura, N., Atsumi, S., Tsuchida, K., Maekawa, H. and Sato, R. (2004). A carotenoid-binding protein (CBP) plays a crucial role in cocoon pigmentation of silkworm (Bombyx mori) larvae. *FEBS Lett.* **567**, 175–178.

Tanne, E. and Sela, I. (2005). Occurrence of a DNA sequence of a non-retro RNA virus in a host plant genome and its expression: evidence for recombination between viral and host RNAs. *Virology* **332**, 614–622.

Taylor, D. J., Leach, R. W. and Bruenn, J. (2010). Filoviruses are ancient and integrated into mammalian genomes. *BMC Evol. Biol.* **10**, 193.

Teixeira, L., Ferreira, A. and Ashburner, M. (2008). The bacterial symbiont Wolbachia induces resistance to RNA viral infections in Drosophila melanogaster. *PLoS Biol.* **6**, e2.

Tian, H., Peng, H., Yao, Q., Chen, H., Xie, Q., Tang, B. and Zhang, W. (2009). Developmental control of a lepidopteran pest Spodoptera exigua by ingestion of bacteria expressing dsRNA of a non-midgut gene. *PLoS One* **4**, e6225.

Timmons, L. and Fire, A. (1998). Specific interference by ingested dsRNA. *Nature* **395**, 854.

Tomari, Y., Matranga, C., Haley, B., Martinez, N. and Zamore, P. D. (2004). A protein sensor for siRNA asymmetry. *Science* **306**, 1377–1380.

Tomari, Y., Du, T. and Zamore, P. D. (2007). Sorting of Drosophila small silencing RNAs. *Cell* **130**, 299–308.

Tomoyasu, Y., Miller, S. C., Tomita, S., Schoppmeier, M., Grossmann, D. and Bucher, G. (2008). Exploring systemic RNA interference in insects: a genome-wide survey for RNAi genes in Tribolium. *Genome Biol.* **9**, R10.

Tsitsiou, E. and Lindsay, M. A. (2009). microRNAs and the immune response. *Curr. Opin. Pharmacol.* **9**, 514–520.

Turner, C. T., Davy, M. W., MacDiarmid, R. M., Plummer, K. M., Birch, N. P. and Newcomb, R. D. (2006). RNA interference in the light brown apple moth, Epiphyas postvittana (Walker) induced by double-stranded RNA feeding. *Insect Mol. Biol.* **15**, 383–391.

Tzou, P., Ohresser, S., Ferrandon, D., Capovilla, M., Reichhart, J. M., Lemaitre, B., Hoffmann, J. A. and Imler, J. L. (2000). Tissue-specific inducible expression of antimicrobial peptide genes in Drosophila surface epithelia. *Immunity* **13**, 737–748.

Tzou, P., De Gregorio, E. and Lemaitre, B. (2002). How Drosophila combats microbial infection: a model to study innate immunity and host-pathogen interactions. *Curr. Opin. Microbiol.* **5**, 102–110.

Umbach, J. L. and Cullen, B. R. (2009). The role of RNAi and microRNAs in animal virus replication and antiviral immunity. *Genes Dev.* **23**, 1151–1164.

Vagin, V. V., Sigova, A., Li, C., Seitz, H., Gvozdev, V. and Zamore, P. D. (2006). A distinct small RNA pathway silences selfish genetic elements in the germline. *Science* **313**, 320–324.

Vaistij, F. E., Jones, L. and Baulcombe, D. C. (2002). Spreading of RNA targeting and DNA methylation in RNA silencing requires transcription of the target gene and a putative RNA-dependent RNA polymerase. *Plant Cell* **14**, 857–867.

Vallier, A., Vincent-Monegat, C., Laurencon, A. and Heddi, A. (2009). RNAi in the cereal weevil Sitophilus spp: systemic gene knockdown in the bacteriome tissue. *BMC Biotechnol.* **9**, 44.

Vannini, L., Ciolfi, S., Dallai, R., Frati, F., Hoffmann, K. H. and Meyering-Vos, M. (2010). Putative-farnesoic acid O-methyltransferase (FAMeT) in medfly reproduction. *Arch. Insect Biochem. Physiol.* **75**, 92–106.

van Rij, R. P., Saleh, M. C., Berry, B., Foo, C., Houk, A., Antoniewski, C. and Andino, R. (2006). The RNA silencing endonuclease Argonaute 2 mediates specific antiviral immunity in Drosophila melanogaster. *Genes Dev.* **20**, 2985–2995.

Voinnet, O. (2005). Non-cell autonomous RNA silencing. *FEBS Lett.* **579**, 5858–5871.

Voinnet, O., Vain, P., Angell, S. and Baulcombe, D. C. (1998). Systemic spread of sequence-specific transgene RNA degradation in plants is initiated by localized introduction of ectopic promoterless DNA. *Cell* **95**, 177–187.

Volpi, S., Bongiorni, S. and Prantera, G. (2007). HP2-like protein: a new piece of the facultative heterochromatin puzzle. *Chromosoma* **116**, 249–258.

Wang, X. H., Aliyari, R., Li, W. X., Li, H. W., Kim, K., Carthew, R., Atkinson, P. and Ding, S. W. (2006). RNA interference directs innate immunity against viruses in adult Drosophila. *Science* **312**, 452–454.

Wang, X. B., Wu, Q., Ito, T., Cillo, F., Li, W. X., Chen, X., Yu, J. L. and Ding, S. W. (2010). RNAi-mediated viral immunity requires amplification of virus-derived siRNAs in Arabidopsis thaliana. *Proc. Natl. Acad. Sci. USA* **107**, 484–489.

Watanabe, T., Totoki, Y., Toyoda, A., Kaneda, M., Kuramochi-Miyagawa, S., Obata, Y., Chiba, H., Kohara, Y., Kono, T., Nakano, T., Surani, M. A., Sakaki, Y., *et al.* (2008). Endogenous siRNAs from naturally formed dsRNAs regulate transcripts in mouse oocytes. *Nature* **453**, 539–543.

Whyard, S., Singh, A. D. and Wong, S. (2009). Ingested double-stranded RNAs can act as species-specific insecticides. *Insect Biochem. Mol. Biol.* **39**, 824–832.

Wilkins, C., Dishongh, R., Moore, S. C., Whitt, M. A., Chow, M. and Machaca, K. (2005). RNA interference is an antiviral defence mechanism in Caenorhabditis elegans. *Nature* **436**, 1044–1047.

Wilson, J. A. and Richardson, C. D. (2005). Hepatitis C virus replicons escape RNA interference induced by a short interfering RNA directed against the NS5b coding region. *J. Virol.* **79**, 7050–7058.

Winston, W. M., Molodowitch, C. and Hunter, C. P. (2002). Systemic RNAi in C. elegans requires the putative transmembrane protein SID-1. *Science* **295**, 2456–2459.

Winston, W. M., Sutherlin, M., Wright, A. J., Feinberg, E. H. and Hunter, C. P. (2007). Caenorhabditis elegans SID-2 is required for environmental RNA interference. *Proc. Natl. Acad. Sci. USA* **104**, 10565–10570.

Wu, Q., Luo, Y., Lu, R., Lau, N., Lai, E. C., Li, W. X. and Ding, S. W. (2010). Virus discovery by deep sequencing and assembly of virus-derived small silencing RNAs. *Proc. Natl. Acad. Sci. USA* **107**, 1606–1611.

Xi, Z., Ramirez, J. L. and Dimopoulos, G. (2008). The Aedes aegypti toll pathway controls dengue virus infection. *PLoS Pathog.* **4**, e1000098.

Xie, Z., Fan, B., Chen, C. and Chen, Z. (2001). An important role of an inducible RNA-dependent RNA polymerase in plant antiviral defense. *Proc. Natl. Acad. Sci. USA* **98**, 6516–6521.

Xiong, C., Levis, R., Shen, P., Schlesinger, S., Rice, C. M. and Huang, H. V. (1989). Sindbis virus: an efficient, broad host range vector for gene expression in animal cells. *Science* **243**, 1188–1191.

Xu, W. and Han, Z. (2008). Cloning and phylogenetic analysis of sid-1-like genes from aphids. *J. Insect Sci.* **8**, 1–6.

Zaidman-Remy, A., Herve, M., Poidevin, M., Pili-Floury, S., Kim, M. S., Blanot, D., Oh, B. H., Ueda, R., Mengin-Lecreulx, D. and Lemaitre, B. (2006). The Drosophila amidase PGRP-LB modulates the immune response to bacterial infection. *Immunity* **24**, 463–473.

Zambon, R. A., Vakharia, V. N. and Wu, L. P. (2006). RNAi is an antiviral immune response against a dsRNA virus in Drosophila melanogaster. *Cell. Microbiol.* **8**, 880–889.

Zhao, M., Soderhall, I., Park, J. W., Ma, Y. G., Osaki, T., Ha, N. C., Wu, C. F., Soderhall, K. and Lee, B. L. (2005). A novel 43-kDa protein as a negative regulatory component of phenoloxidase-induced melanin synthesis. *J. Biol. Chem.* **280**, 24744–24751.

RNAi Unravels the Biology of the Hemimetabolous and Ametabolous Insects

Darron A. Cullen[1]

School of Biological Sciences, The University of Sydney, Sydney, Australia

Abstract

RNA interference (RNAi) has proven to be among the most important developments in molecular biology-driven studies of insects over the last decade. Using exogenous double-stranded RNA to generate post-transcriptional knockdown phenotypes, this technique enables the functional characterization of an ever-increasing set of genetic sequence data.

Whilst routinely employed in studies of model holometabolous insects like *Drosophila melanogaster* and *Tribolium castaneum*, RNAi has also facilitated research into the hemimetabolous and ametabolous orders. This looks set to increase with the recent publication of the first genomic datasets for insects without full metamorphosis; the pea aphid, *Acyrthosiphon pisum*, and the human body louse, *Pediculus humanus humanus*.

In anticipation of a heightened interest in this field, this review surveys the diverse range of recent research employing RNAi in ametabolous and hemimetabolous insects and, in doing so, demonstrates the broad utility of the technique for future studies of development, physiology and behaviour.

[1] Current address: Department of Zoology, University of Cambridge, Cambridge, United Kingdom.

ADVANCES IN INSECT PHYSIOLOGY VOL. 42
ISBN 978-0-12-387680-5
DOI: 10.1016/B978-0-12-387680-5.00002-1

1 Introduction

Post-transcriptional gene silencing by double-stranded RNA (dsRNA) has revolutionized the field of molecular biology since its definitive discovery in the nematode *Caenorhabditis elegans* (Fire *et al*., 1998; Timmons and Fire, 1998), following previous insights from plants and *Neurospora* (Jorgensen, 1990; Romano and Macino, 1992). Key components of the RNA interference (RNAi) pathway are the Dicer enzyme and the RNA-induced silencing complex (RISC), which are thought to have evolved a dual purpose: the recognition and degradation of foreign dsRNA, acting as a defence against viral genomes and transposable elements (Stram and Kuzntzova, 2006; Voinnet, 2001); and internal molecular control mediated by endogenous microRNAs, whereby mRNAs can be sequestered before translation, particularly during development (Wang and Li, 2007; Zhao and Srivastava, 2007).

Over the past decade, exploitation of the Dicer-RISC machinery with exogenous dsRNA has become an invaluable tool in the characterization of genetic sequences, through the down-regulation of sequence-sharing mRNAs. This technique has proven particularly fruitful since its discovery coincided with the publication of several key genomes, including those of *C. elegans* and some insects, particularly the model fruit flies *Drosophila melanogaster* and *Drosophila pseudoobscura* (Diptera). Work in *D. melanogaster* was the first to demonstrate that insects are amenable to transcript suppression through RNAi (Kennerdell and Carthew, 1998). This enabled the development of high-throughput screening methods for uncharacterized genetic sequences in *Drosophila*, and the establishment of an online FlyRNAi database (Flockhart *et al*., 2006). Full genome sequences have also been annotated for the mosquitoes *Anopheles gambiae* and *Aedes aegypti* (Diptera), the honeybee *Apis mellifera* (Hymenoptera), the parasitoid jewel wasp *Nasonia vitripennis* (Hymenoptera) and the red flour beetle *Tribolium castaneum* (Coleoptera).

The wealth of genetic data available for these holometabolous insects has greatly facilitated research into their biology, with RNAi now commonplace in the study of all of them. However, there remains a notable lack of sequence information for the insects that do not undergo full metamorphosis, broadly grouped into ametabolous and hemimetabolous taxa (or apterygote and exopterygote taxa). In spite of this limitation, functional genomic investigation into the biology of these animals is becoming increasingly fruitful, and this looks set to continue exponentially with the recent publication of the first fully sequenced and annotated genome of a hemimetabolous insect, the pea aphid *Acyrthosiphon pisum* (International Aphid Genomics Consortium, 2010). This was followed only a few months later by the genome of the human body louse, *Pediculus humanus humanus* (International Louse Genomics Consortium, 2010).

In light of these recent developments, and anticipating a surge of interest in these lineages, this review seeks to consolidate the literature describing

RNAi-driven investigation into the hemimetabolous and ametabolous insects and, in doing so, highlight the utility of the technique in answering a broad range of biological questions.

2 Growth and maintenance of the insect body plan

2.1 EMBRYONIC DEVELOPMENT

The overall body plan of insects is highly conserved across all orders, but the mode of segmentation during embryonic development varies widely. These divergent mechanisms have historically been categorized into three categories: 'short', 'intermediate' and 'long' germ-band embryogenesis. Short and intermediate germ-band embryogenesis is widespread throughout the insect world, including in all the hemimetabolous insects and the model beetle *Tribolium castaneum*. During this mode of development, only the anterior body segments (up to seven forming the head, and three the thorax) are determined during the blastoderm stage of development, with abdominal segments (8–11 total) forming later, in a successive manner from a posterior growth zone. By contrast, in long germ-band embryogenesis all segments are specified simultaneously at the blastoderm stage. This mode is confined to some of the holometabolous species (including *Drosophila*), and is thus thought to have evolved from the short/intermediate germ modes (Liu and Kaufman, 2005b; Mito *et al.*, 2006).

Kaufman and colleagues pioneered the systematic use of RNAi in the hemimetabolous insects, to analyse the roles of some well-characterized *Drosophila* genes in segmentation and appendage formation in the intermediate germ-band milkweed bug, *Oncopeltus fasciatus* (see Angelini and Kaufman, 2005b; Hughes and Kaufman, 2000). This has enabled them to build a comparative framework of loss-of-function phenotypes, through which to investigate the divergent evolutionary development between the two modes of development. [They are often mistakenly accredited with being the first group to successfully perform RNAi in a hemimetabolous insect (e.g. Bellés, 2010), but this is not actually the case; see Marie *et al.*, 2000]. In addition, work by Noji and Mito has concurrently unravelled some functions of the same (and related) genes in their orthopteran model, the two-spotted cricket *Gryllus bimaculatus*. These research programs have been reviewed in great detail elsewhere (see Angelini and Kaufman, 2005b; Liu and Kaufman, 2005b; Mito *et al.*, 2010, 2011; Nakamura *et al.*, 2010) but some key findings are outlined here, alongside further studies of hemimetabolous insect development employing RNAi.

In the first use of RNAi in *Oncopeltus*, Hughes and Kaufman (2000) investigated the interactive role of three Hox genes (*Deformed, proboscipedia* and *Sex combs reduced*) in the specification of the bug's highly specialized sucking mouthparts. They discovered that, while all are necessary for the development of the gnathal appendages in this insect, the mode of action differs from that known in the holometabolous insects *Drosophila* and *Tribolium*. Microinjector-delivered dsRNA of all three genes in synchrony caused all of the feeding appendages to develop as antennae. By contrast, RNAi against *proboscipedia* alone transformed the embryonic labium into a thoracic leg, complete with comb and claws, while knockdown of *Sex combs reduced* (*Scr*) altered the same gnathal structure towards an appendage of mixed identity. Angelini *et al.* (2005) went on to show that *proboscipedia* and *Scr* function cooperatively in specifying the identity of the labium in *Oncopeltus*, in contrast to their epistatic interaction in *Drosophila* in which *Scr* activates *proboscipedia*.

This canon of work was taken a step further by Popadić and co-workers, who showed that *Scr* knockdown did not affect labial identity in third, fourth or fifth instar nymphs of *Oncopeltus*, highlighting divergent functions of the gene between embryonic and post-embryonic development (Chesebro *et al.*, 2009). RNAi against *Scr* at these later life stages did, however, lead to the development of wings on the pronotum, signalling a commonality with the holometabolous insects wherein *Scr* plays a role in wing repression at this segment (Chesebro *et al.*, 2009). Popadić's group found a similar result in the American cockroach, *Periplaneta americana*. *Scr* action appeared to be limited to the head during embryogenesis (in which *Scr*-RNAi led to labial palps with a row of bristles and a large femur spur), while knockdown at various nymphal stages predominantly affected the prothoracic segment (Hrycaj *et al.*, 2010). As with *Oncopeltus*, *Scr*-knockdown animals developed a misshaped pronotum over successive moults, which more closely resembled the notum of the adjacent segment (T2) of wild-type adults. The strongest knockdown phenotype was the development of ectopic wings on the pronotum, again similar to that found in *Oncopeltus* and leaving exciting implications for the study of insect wing evolution (Hrycaj *et al.*, 2010). However, in contrast to *Oncopeltus*, *Scr* in *Periplaneta* also proved to be critical in the formation of the dorsal ridge, a structure separating the head and thorax in all insects; knockdown embryos developed an additional segment between the head and prothoracic segments.

Further innovation by Kaufman's group came via the investigation of the *Oncopeltus* orthologues of two 'gap genes', which divide the early embryos into broad regions prior to their differentiation into the specific segments of the body. These studies of *hunchback* and *Krüppel* utilized parental RNAi for the first time in a hemimetabolous insect (Liu and Kaufman, 2004a). By injecting dsRNA into virgin females prior to mating, the authors found a much higher penetrance than in direct-injected embryos. *hunchback* was found to be necessary for suppressing abdominal identity in the developing head and thoracic segments, as well as controlling later germ-band growth in the abdomen. Concurrent work used RNAi against *Krüppel*, resulting in the loss of several segments within the thorax and abdomen, showing it to be necessary during both the blastoderm and germ-band stages of short/intermediate germ-band insect development. In these respects, RNAi phenotypes for *hunchback* and *Krüppel* in *Oncopeltus* were similar to the homologous mutant phenotypes in

Drosophila (Liu and Kaufman, 2004b), and were later seen again in the cricket. *hunchback* knockdown in *Gryllus* led to the expression of abdominal identity in the head and thoracic tagmata, while RNAi against *Krüppel* caused disruption of development across the entire embryo (Mito *et al.*, 2005, 2006). The role of *hunchback* does not appear to be universal, however; parental RNAi in the migratory locust, *Locusta migratoria*, yielded even more severely deleterious phenotypes than those reported in *Gryllus* and *Oncopeltus* (He *et al.*, 2006, 2010). Only the pre-gnathal segments of the head were unaffected in some embryos, while affected segments did not appear to take on abdominal identity, suggesting a divergent epistatic pathway of *hunchback* within the larger orthopteran. Parental RNAi was also performed against *Oncopeltus wingless*, which uncovered a role during body segmentation, but not appendage development (Angelini and Kaufman, 2005a). By contrast, depletion of *wingless* via microinjection of dsRNA into *Gryllus* embryos did not result in a detectable knockdown phenotype (Miyawaki *et al.*, 2004). The authors speculate that this could be due to a redundant effect of other sequences in the *wingless/Wnt* family, of which seven are known in *Drosophila* (Nakamura *et al.*, 2007) and nine have been identified in *Tribolium* (Bolognesi *et al.*, 2008).

More recent work in *Oncopeltus* has focused on *giant* and *knirps*, two additional gap genes. Liu and Patel (2010) investigated the role of *giant* in the milkweed bug, since it functions as a gap gene during segmentation of segments 5–7 in *Drosophila* but not in the short germ-band beetle *Tribolium* (Bucher and Klingler, 2004; Kraut and Levine, 1991). Parental RNAi resulted in embryos lacking the labium and maxillary appendages, with severely stunted abdominal segmentation. This confirms the role of *giant* as a gap gene in *Oncopeltus*, and places the gap function as ancestral for holometabolous insects; this function has seemingly been lost in the *Tribolium* lineage. Ben-David and Chipman (2010) highlighted a difference between holometabolous and hemimetabolous insects, by using parental RNAi to further characterize *giant* expression during blastoderm patterning. Through investigating its interactions with *Krüppel*, *hunchback* and *knirps*, they discovered that the mutual repression between all four genes is much less pronounced than in *Drosophila*. The embryonic subdivision mediated by these gap genes is controlled further in *Drosophila* by pair-rule genes, which are only expressed in every other segment of the embryo; one such gene that is well characterized in *Drosophila* is *even-skipped*. By contrast, Liu and Kaufman (2005a) found that *even-skipped* is not a pair-rule gene in *Oncopeltus* at all. Instead, direct and parental RNAi against the sequence created huge defects across the entire body plan, with only the eye and antennae recognizable in some cases. Furthermore, staining for the protein products of *Krüppel* and *hunchback* suggested that *even-skipped* regulates the expression of these gap genes, implicating *even-skipped* in multiple stages of embryonic development in *Oncopeltus*. Mito *et al.* (2007) reported similarly deleterious effects of *even-skipped* RNAi in *Gryllus* embryogenesis, and also found that the gene had pair-rule-like functions in the developing cricket.

Mito and Noji's group have also investigated several genes for which RNAi-based experiments have yet to be published in *Oncopeltus*. In their initial cricket RNAi-based experiment, they silenced the *Gryllus* orthologue of the *Drosophila* segment-polarity gene *armadillo*, via microinjection of dsRNA into developing eggs (Miyawaki *et al.*, 2004). The resulting phenotype was a lack of abdomen and abnormal anterior segments. Simultaneous attempts to silence *hedgehog* (and *wingless*, as mentioned above) were unsuccessful, with a barely noticeable alteration of expression levels in immuno-stained whole-mount embryos and no noticeable change in morphology; however, their numbers were very low ($n=3$ for some treatments). Nevertheless, Noji's group were still able to postulate the Hox gene expression pattern during early embryogenesis in *Gryllus*, a model that they have added to and modified throughout much of their subsequent work. This includes parental RNAi against the transcription factor *caudal*, which also produced a marked effect on the embryo phenotype, lacking gnathum, thorax and abdomen; only the anterior head remained in some cases (Shinmyo *et al.*, 2005). *caudal* encodes a transcription factor, which is co-expressed with the gap genes *hunchback* and *Krüppel*, suggesting that it is these genes that *caudal* is up-regulating. Shinmyo *et al.* (2006) went on to show that *brachyenteron*, which is necessary during differentiation of the posterior gut and Malpighian tubules in *Drosophila*, has a highly converved function in *Gryllus*. Malformation of these digestive structures in response to *brachyenteron* down-regulation in *Gryllus* led to death by starvation in the first instar. However, *brachyenteron* did not appear to play a role in abdominal segmentation. Parental RNAi has also been used to investigate embryonic segmentation in the developing short germ-band cockroach *Periplaneta*, which was shown to be dependent on the *Notch* signalling pathway (Pueyo *et al.*, 2008). Knockdown of *Notch* and *hairy* showed that expression of these genes (along with *Delta*) is necessary for the laying down of new boundaries between abdominal segments. This is similar to *Notch* function in vertebrate somite embryogenesis, and the conserved rhythmic nature of gene expression led the authors to suggest that this developmental pathway might have evolved in a common ancestor to insects and vertebrates, as opposed to having evolved convergently (De Robertis, 2008; Pueyo *et al.*, 2008).

The specification of appendages and feeding apparatus is a further avenue of research to employ RNAi. In some of their earlier work in the field, Kaufman and colleagues silenced the *Oncopeltus* orthologues of the leg gap genes *dachshund*, *homothorax* and *Distal-less* (Angelini and Kaufman, 2004). These genes were already known to be conserved in their expression patterns during leg formation between *Oncopeltus* and *Drosophila*, but their functions were as yet unresolved in the milkweed bug. *dachshund* was shown to be necessary for development of the mandibular and maxillary feeding stylets in *Oncopeltus*, while *homothorax* RNAi affected both stylet and labial identity. Interestingly, *homothorax*-RNAi also produced severely deleterious phenotypes in *Gryllus*, in which all head appendages differentiated with a mixed identity of legs and antennae (Ronco *et al.*, 2008). By contrast, RNAi showed that *Distal-less* is not required in embryonic specification of antennal identity in *Oncopeltus*, though it did appear to play a role in antennal development post-identification. In the gnathal appendages, *Distal-less* RNAi did not affect the development of the stylets, but the distal labial segment was deleted. Further investigation of *Distal-less* function utilized the firebrat, *Thermobia domestica* (Thysanura), an apterygote insect more distantly related to *Drosophila* than the hemimetabolous insects are. Among the most noticeable anatomical features of the firebrat is the caudal filament, which protrudes from the posterior-most segment of the abdomen and can be as long again as the main body of the insect. Flanked on each side by two smaller cerci, the trio of structures is thought to aid in predator avoidance and is also found within some of the hemimetabolous insect orders, particularly the Orthoptera (Edwards and Reddy, 1986). Rogers *et al.* (2002) used antibody staining in developing firebrat embryos to highlight an accumulation of the Distal-less transcription factor in all three caudal structures, as well as considerable expression distally within the developing legs and gnathal appendages. Ohde *et al.* (2009) went on to disrupt this pathway with RNAi against *Distal-less*, and proved that this gene is indeed required to form the caudal filament, thoracic appendages, and feeding apparatus. Introduced via microinjection into the developing embryo, dsRNA-knockdown animals exhibited a shortened caudal filament, while only a small deletion of the distal tip was observed in the cerci. The developmental disruption of the caudal filament in particular has perhaps proven the most insightful since, in contrast to the associated cerci, the caudal filament is not thought to be a true appendage; this work suggests a common developmental pathway nevertheless (Rogers *et al.*, 2002).

More recently, appendage development in *Oncopeltus* has been unravelled further. Herke *et al.* (2005) injected dsRNA for the Hox gene *Antennapedia*, which specifies leg identity over the default antenna-like appendage in *Drosophila*. It appears to play no such role in *Oncopeltus*, though it was still necessary for leg segmentation. By contrast, RNAi against *tiptop* demonstrated its necessity for both leg identity and development; knockdown animals grew antenna-like structures at the distal end of their thoracic legs. Mahfooz *et al.* (2007) confirmed a hypothesis based on expression data (Mahfooz *et al.*, 2004) that the Hox gene *Ultrabithorax* is important in growth of the hind legs. The authors suggest that an evolutionary modification of *Ultrabithorax* expression has given rise to larger jumping femurs in other insects (which they further demonstrated through *Ultrabithorax* RNAi in the house cricket, *Acheta domestica*). Interestingly, work using parental RNAi in another hemipteran, the water strider *Gerris buenoi*, suggests a divergent role of *Ultrabithorax* during development of the particularly long limbs of this species (Khila *et al.*, 2009). Work in *Oncopeltus* went on to demonstrate that *nubbin* suppresses leg growth along the abdominal segments, with knockdown embryos developing bizarre ectopic abdominal appendages as well as an altered morphology in the head and thoracic legs (Hrycaj *et al.*,

2008). This work suggests that *nubbin* acts upon the Hox genes *Ultrabithorax* and *abdominal-A* in *Oncopeltus*, a role it does not share with *Drosophila nubbin*. It appears, therefore, that variation exists in the genetic mechanisms maintaining a limbless abdomen, a feature that originated early in the evolutionary history of insects.

The egg gives rise not only to the embryo, but also to extraembryonic membranes that protect and nourish the developing insect. The genetic control of extraembryonic development has been relatively understudied in insects (Panfilio, 2008), and this is largely due to the fact that the extraembryonic components are much reduced in *Drosophila*, forming a single membrane known as the amnioserosa. However, most insects retain a separate amnion and serosa, and Panfilio (e.g. Panfilio *et al.*, 2006) has been instrumental in using the milkweed bug to unravel the genetic control of extraembryonic development. In particular, she has focused on the ametabolous- and hemimetabolous-specific process of blastokinesis, through which the embryo and two extraembryonic membranes move around inside the egg during development (Panfilio and Roth, 2010). Parental RNAi has shown that the *Hox3* gene (*zen*) is crucial during katatrepsis, the latter part of blastokinesis in which the embryo emerges from the yolk and alters its orientation within the egg. Knockdown individuals remained in the yolk and completed their development inside out (Panfilio, 2009; Panfilio *et al.*, 2006). Whilst still in its early stages, this research looks set to decode many key evolutionary events in the field of insect development in the near future, and Panfilio (2008) provides a more comprehensive review of the topic.

Studies of embryonic development highlight, perhaps better than any other biological discipline, the utility of RNAi in studies of comparative gene function and evolution. Prior to RNAi, the site of expression for a given gene could be determined using in situ hybridization and antibody staining of the protein, while the extent of its expression was measurable at the protein level using ELISA-based techniques, or at the transcript level using RT-qPCR. However, it is only with RNAi that gene function itself can be easily elucidated in a systematic way, via the generation of loss-of-function mutants. Moreover, the cross-taxa comparisons that this facilitates ensures that all genes can be considered not only from a developmental context, but an evolutionary one as well, as demonstrated above for *hunchback* (He *et al.*, 2006), *even-skipped* (Liu and Kaufman, 2005a) and *Distal-less* (Angelini and Kaufman, 2004; Ohde *et al.*, 2009), among others.

2.2 POST-EMBRYONIC DEVELOPMENT

The process of moulting is key to the incremental development of the hemimetabolous insects, in which periods of growth in developing nymphs are punctuated with periodic ecdysis events, ensuring that internal expansion is not hindered by the relatively inelastic properties of the cuticle. Moulting in the hemimetabolous insects is probably best understood in the German cockroach *Blattella germanica*, thanks to work by Bellés and co-workers (Cruz *et al.*, 2006; Martín *et al.*, 2006).

They used RNAi to scrutinize the ecdysteroid-driven pathways involved in moulting, having been the first group to successfully test RNAi in this small species of cockroach (Ciudad *et al.*, 2006). They began by individually silencing each subunit of the heterodimeric ecdysone receptor, *EcR* and the *RXR*-homologue *Ultraspiracle* (Cruz *et al.*, 2006; Martín *et al.*, 2006). RNAi against either sequence led to identical nymphal phenotypes, which were similar to the 'double mouth-hooks' phenotype reported in ecdysteroid-knockout *Drosophila*. Duplication of cuticular structures occurred within many parts of the cockroach body, including a double-mandible and the old tracheal cuticle visible within the lumen of the new one. The result was an incomplete shedding of the exuvia (ecdysis) and a fatal arresting of the moulting process. Post-mortem analysis of these insects revealed that they had successfully laid down a new cuticle and undergone apolysis (separation of the old cuticle from the epidermis) to the same extent as control animals. The problematic step for both cohorts of knockdown insects was the shedding of the old nymphal cuticle, that is, the ecdysis behaviour. Ecdysis is dependent on a precise sequence of body contractions, initiated and mediated by various peptide signals which, the authors conclude, must be synthesised and/or released downstream of the binding of ecdysone to its receptor. Subsequent work sought to elucidate further steps along this signalling cascade and focused on a handful of nuclear receptors known to be involved in moulting and metamorphosis in *Drosophila*. Cruz *et al.* (2007) cloned three isoforms of the *hr3* receptor homologue and used RNAi against all three in synchrony to produce a deleterious phenotype identical to that obtained from the knockdown of *EcR* or *RXR/USP*. Interestingly, knockdown of any single isoform on its own did not provoke a deleterious effect on ecdysis, suggesting that the different transcripts have redundant functions during this part of the life cycle. Similar work with the nuclear receptor gene *ftz-f1* produced the same pattern of arrested ecdysis in response to RNAi-mediated silencing, and also showed that expression of this transcript follows, and is tightly regulated by, expression of the *hr3* receptor gene (Cruz *et al.*, 2008). These functions of *hr3* and *ftz-f1* are conserved with respect to their known roles in *Drosophila*, and are thus likely to be ancestral constituents of the moulting pathway.

Further insights into the ecdysis process itself have come from the orthopteran *Locusta migratoria*. Ecdysis can only occur after enzymatic breakdown of the old cuticle, and the generation of a new chitin exoskeleton underneath that is sufficiently strong to break through. Wei *et al.* (2007) isolated an mRNA transcript encoding a novel protease from within the interstitial moulting fluid of developing locusts, before using RNAi to prove its necessity to the cuticle-recycling process. Knockdown of this transcript, deemed by sequence analysis to belong to the trypsin-like serine protease family, produced clear defects in locusts moulting from fourth to fifth instar, and from fifth instar to adult. Approximately one third of the knockdown animals did not experience sufficient cuticle degradation to moult successfully and died within 72 h of ecdysis arrest. A smaller proportion of the treated animals were able to moult successfully, but the shed exuviae were measurably thicker than those produced by wild-type individuals. More recently, Zhang *et al.* (2010) investigated the function of two alternative splicing-derived variants of the *chitin synthase 1* (*CH1A* and *CH1B*) gene in *Locusta* with RNAi. Concurrent knockdown of both variants in second instar nymphs resulted in two main classes of deleterious phenotype: one in which ecdysis was unsuccessful, as the new cuticle was not sufficiently strong to break through the old exoskeleton; and one in which the animals moulted successfully, but then failed to walk due to a lack of mechanical support by the cuticular architecture, which could not sustain their body weight. Both phenotypes proved fatal and, since chitin is not found in plants or vertebrates, the authors suggest that this pathway could be the target of a novel means of crop pest management.

The incremental nature of hemimetabolous insect development dictates that growth and maturation occur simultaneously and are thus synchronized more closely than in their metamorphosing counterparts in which growth occurs (primarily) within the larval stages, prior to a distinct maturation stage at metamorphosis. Mané-Padrós *et al.* (2008) showed that a homologue of the *Drosophila E75* nuclear hormone receptor plays just such a role in this synchronization process in *Blattella*, and Erezyilmaz *et al.* (2009b) also demonstrated a function of *E75* in *Oncopeltus*. Silencing the *E75* transcript in *Blatella* did not cause a fatally arrested moult; in fact, knockdown final-instar nymphs for *E75* did not moult at all and persisted in their final immature stadium for up to 90 days. These animals continued to feed to the point that their abdomens became noticeably distended, whereas control animals at the same stage only fed for the usual 4 days before moulting to adulthood. Upon closer inspection, these animals' failure to enter the apolytic process was found to be due to a premature degeneration of the prothoracic gland, which synthesizes the ecdysteroid hormones. This does not normally occur until adulthood and, remarkably, other adult-specific aspects of development also continued underneath the morphologically immature cuticle. Of particular note was the reproductive cycle, including the synthesis of yolk proteins by the fat body and their subsequent incorporation into developing oocytes. *E75* therefore appears to link moulting to developmental progression in these insects and, interestingly, there is a similar level of functional redundancy between five isoforms of *E75* as was noted above for the three versions of *HR3*. Further investigation into the degradation of the prothoracic gland showed that two proteins maintain tissue viability in this structure during the nymphal stadia (Mané-Padrós *et al.*, 2010). Silencing these 'inhibitors of apoptosis' (IAPs) with RNAi showed that one in particular is essential for preventing post-embryonic tissue degeneration across all internal organs that were assayed (including the prothoracic gland), with knockdown animals dying within 48h of treatment.

The holometabolous life history, comprising a larval stage that is morphologically distinct from the adult stage, is argued to be one of the major factors contributing to the broad speciation of this superorder of insects (McGavin and Lewington, 2001). While not the focus of this review, these species are thought to have evolved from within the hemimetabolous orders, and much of our current understanding of this evolutionary change in life cycle comes from the work of Truman and Riddiford (e.g. Truman and Riddiford, 1999, 2007). With the emergence of RNAi they have been able to investigate the evolutionary origins of metamorphosis from within the hemimetabolous insect lineage and focused first on the expression of the transcription factor *broad* in the milkweed bug *Oncopeltus fasciatus broad* specifies larval-pupal development within the metamorphosing insects *Drosophila* and *Manduca sexta*, and this is the sole process that it appears to regulate within those species (Zhou and Riddiford, 2002). By contrast, *broad* appears to be involved in embryonic development in *Oncopeltus* and continues to be expressed at each nymphal moult until the final shedding to adulthood (Erezyilmaz *et al.*, 2006). RNAi against *broad* in third-instar nymphs did not arrest the moulting process itself, but instead affected the heteromorphic transitions typical of the incrementally developing hemimetabolous insects, as fourth-instar nymphs retained the pigmentation and wing-bud formation from their previous stadium. Parental RNAi against *broad* produced a wide range of deleterious embryological phenotypes in *Oncopeltus* (Erezyilmaz *et al.*, 2009a) and also implicated the *broad* gene in development of *Blattella* (Piulachs *et al.*, 2010). While the function of the gene is still uncertain, *broad* clearly appears to support progressive change during moulting of the hemimetabolous insects, and Erezyilmaz *et al.* (2006) speculate that complete metamorphosis arose when its expression became confined to a single instar: the prepupal stadium of the Holometabola.

The moulting process affords nymphs of hemimetabolous insects the ability to regenerate amputated limbs in an incremental fashion, between successive nymphal stadia. This process, also known as intercalary regeneration, has been the focus of much work in *Gryllus* by Noji and Mito (Nakamura *et al.*, 2008b). Classic experiments (see Anderson and French, 1985) first demonstrated that, after amputation of a limb segment and subsequent reattachment at a different position, new cuticular structures corresponding to the missing segments are laid down during following moults. Nakamura *et al.* (2007) performed similar experiments in *Gryllus* and found that RNAi for some developmental genes would only yield an observable phenotypic change during regeneration and not between nymphal moults of unadulterated insects. They dubbed this phenomenon ‘regeneration-dependent RNAi’ and showed that RNAi targeting Wnt signalling via the segment-polarity gene *armadillo* could prevent regeneration. RNAi against the epidermal growth factor receptor (EGFR) produced a less severe phenotype (Nakamura *et al.*, 2008a). Knockdown animals lacked the tarsus and pretarsus on their regenerated limbs, indicating the necessity of EGFR in the patterning of the distal portion of these appendages. Bando *et al.* (2009) have also shown that RNAi against homologues of the *Drosophila* leg patterning genes *Dachous* and *fat* resulted in undersized regenerated limbs, whereas RNAi against *expanded* or *Merlin* had the opposite effect, inducing an over-proliferation of cells that made the legs longer.

The efficacy of RNAi in *Gryllus* led to the suggestion that the cricket could be adopted as a new model system for understanding human neurological disorders (Hamada *et al.*, 2009). As a substantially larger insect than *Drosophila, Gryllus* is easier to manipulate and, as discussed in more detail in Section 4, has neurons that are more amenable to electrophysiological examination due to their greater size. Hamada *et al.* (2009) used RNAi to silence homologues of two human genes implicated in well-known neurological conditions. Defective dopamine signalling is known to play a role in the onset of Alzheimer's disease and Parkinson's disease, and knockdown of the dopamine receptor *DopR1* in cricket nymphs resulted in morphological defects in their wings, which were not aligned in a regular position upon moulting to adulthood. However, more pronounced phenotypic effects were apparent after silencing the *fmr1* gene, defects in which are known to cause Fragile X syndrome, the most common form of inherited mental disability in humans. *fmr1*-knockdown resulted in a more severe category of morphological defect in the adult wings than was caused by *DopR1*-RNAi, such that it had a significant effect on male calling song. The authors reason that *fmr1* must play a role in wing muscle growth during the imaginal moult, because knockdown of the same gene in adult males did not affect their calls at all; *fmr1* must therefore act during development, and not in an adult behavioural pathway. While this work is interesting in its own right, any link to human neurological disorders from the conferred phenotypes in these studies is arguably tenuous at best.

RNAi has also been used to investigate eye growth across successive nymphal stadia. Dong and Friedrich (2005) silenced the homologue of the ommochrome pigment *vermillion* via microinjection into first-instar nymphs of the grasshopper *Schistocerca americana*. The resulting nymphs expressed a reduced colouration of the eye, which persisted across two instars, for up to 2 weeks, as the animals developed. Further investigation of eye development (Dong and Friedrich, 2010) demonstrated that knockdown of *eyes absent* and *sine oculis* postponed retinal development in *Schistocerca* nymphs, though retinal development resumed once the transient effects of RNAi had worn off.

Gomez-Orte and Bellés (2009) have advanced our understanding of the evolution of metamorphosis by determining that the process is mediated, at least in part, by microRNAs (miRNA) in *Blattella*. This had already been shown in *Drosophila* (Caygill and Johnston, 2008; Sokol *et al.*, 2008) but it was not previously known whether miRNAs play a role in the more ancestral mode of metamorphosis. The enzyme Dicer-1 is a key constituent of the pathway in all animals, and the RNAi method with which this review is concerned is built around the exploitation of its function. Dicer-1 breaks down miRNA precursors into mature miRNAs and also produces the small interfering RNAs (siRNAs) in RNAi. Gomez-Orte and Bellés turned this enzyme against itself in *Blattella*, by introducing dsRNA encoding a portion of the *Dicer-1* transcript. The result was a reduction in the amount of mature miRNAs in sixth-instar nymphs, which would ordinarily moult into the adult stage. Instead, moulting produced a seventh-instar 'nymphoid' phenotype, suggesting that certain miRNAs created by Dicer-1 act to suppress mRNA transcripts expressing various nymphal features.

2.3 TERMITE CASTE DIFFERENTIATION

The moult-led process of termite caste differentiation is a peculiar case of hemimetabolous development. Termites are the best-known hemimetabolous insects that express eusociality, and thus provide an important system through which to investigate this developmental division of labour. While superficially similar to the eusocial systems seen within the Hymenoptera (wasps, bees and ants), a cursory look at any insect phylogeny highlights the considerable evolutionary distance between the two orders, making it apparent that eusociality has evolved independently in each lineage. Recent work has actually placed the evolutionary origins of the termites within the cockroaches (Inward *et al.*, 2007; Lo *et al.*, 2007) and being hemimetabolous affects their eusocial dynamic quite markedly. This is perhaps best understood within the termite species *Reticulitermes flavipes*, in which immature individuals are predominantly recruited into the worker caste. While performing their duties within the colony, which include foraging, tunnelling and brood tending, workers remain highly plastic. Individuals can remain as workers via a succession of 'status quo' moults (Zhou *et al.*, 2006a [see Fig. 1 in that paper]; Zhou *et al.*, 2006b), differentiate into secondary reproductive individuals, or moult into the soldier caste, which arises entirely from within the workers (Korb and Hartfelder, 2008). This contrasts with their better-known holometabolous counterparts in bees and ants, in which workers and soldiers are necessarily adult, and larvae require constant worker attention.

The moult-led morphogenesis between the nymphal stages of worker and soldier, via an intermediate 'pre-soldier' stage, is induced by a two- to fivefold increase in juvenile hormone (JH) within the haemolymph. This pathway has been the focus of work by Scharf and Zhou (Zhou *et al.*, 2006a,b, 2007), who have implicated two hexamerin genes as being important in the worker-soldier transition. Hexamerins are large storage proteins present in all insects, which evolved from the respiratory hemocyanins present across the arthropod phylum (Hagner-Holler *et al.*, 2007) and are known to bind to JH (Braun and Wyatt, 1996). Zhou and Scharf demonstrated that RNAi against *Hex-1* and *Hex-2* within worker *Reticulitermes* promoted development into the soldier caste, suggesting that these hexamerins operate as a caste-regulatory factor by buffering levels of JH post-synthesis. Hexamerins are known to play a diverse range of roles across the insects, including multiple pathways in cuticle formation and hormone transport, and even phase change in locusts (Burmester, 1999; Kang *et al.*, 2004) and thus would be good targets for RNAi-led analyses of further insect species.

In contrast to the hymenopteran social insects, termite colonies often have more than one reproductive individual in a colony. These typically constitute a single queen (or 'primary reproductive'), and a number of other productives arising from within the subordinate nymphal castes; however, the majority of the colony remains reproductively suppressed. Since workers are immature, they must moult if they are to assume a reproductive role within the colony. Korb *et al.* (2009) discovered that removal of a queen from colonies of *Cryptotermes secundus* instigated aggressive behaviour by would-be secondary reproductives towards their fellow workers, in the form of butting. Workers that eventually moulted into a reproductive state displayed a higher level of butting than workers that did not change caste. This reproductive dominance behaviour was subsequently used as a proxy for perceived queenlessness within an RNAi-based experiment. Knockdown of the *Neofem2* gene in eight queens resulted in their colonies displaying a higher level of butting than control colonies. *Neofem2* is a homologue of a communication-related gene in the Maderian cockroach *Leucophaea maderae*, and this work strongly suggests that it also plays a critical role in queen-worker communication in *Cryptotermes*.

3 Reproduction

Reproductive processes within the hemimetabolous insects is a relatively under-represented field addressed via RNAi-knockdown experiments. This is surprising given the efficacy of parental RNAi, in which the effects of gene silencing in adult females can be passed on to their progeny. This has been used to great effect, especially within the study of insect development (see Section 2 of this review), and suggests that the female reproductive organs are permeable to dsRNA. However, only a handful of studies have dealt with the reproductive cycle itself.

Ghanim *et al.* (2007) down-regulated the homologue of *Drosophila chickadee* in ovaries of the hemipteran whitefly *Bemisia tabaci*, which caused a severe disruption in oocyte development. *chickadee* encodes the small actin-binding protein profilin, which helps to transfer cytoplasmic components into the developing egg but, since all knockdown females died within 2 days of treatment, further analysis of *chickadee* function was necessarily limited. This appears to be the only paper that specifically targets the reproductive organs with RNAi in a hemipteran.

By contrast, the cockroach *Blattella germanica* has proven a far more fruitful model for studying the reproductive cycle with RNAi, and Bellés and colleagues have again led the way. Furthermore, this body of work includes their first paper to utilize systemic RNAi successfully in this insect. Ciudad *et al.* (2006) injected dsRNA for the adult-specific vitellogenin receptor gene into the haemolymph of mature females and demonstrated an effective knockdown of the transcript using Northern blot analysis. Vitellogenins are yolk proteins produced in the fat body, which are incorporated into the maturing oocyte via receptor-mediated endocytosis. Knockdown of the receptor led to low yolk content, and a high vitellogenin concentration in the haemolymph, demonstrating the necessity of this receptor for the process of oogenesis. Similar work used RNAi to knock down the receptor for lipophorin. Lipophorin levels decreased in the ovaries, showing that the pathway mediated by this lipid transport protein is also necessary for the developing *Blattella* eggs (Ciudad *et al.*, 2007).

Work by Bellés and colleagues investigated the role of allatostatin in the reproductive cycle. This neuropeptide was previously shown to regulate JH synthesis by the corpora allata in another species of cockroach, *Diploptera punctata*, and was named after the organ as a result of this discovery (Pratt *et al.*, 1989). Maestro and Bellés (2006) performed RNAi against the *Blattella* homologue of this gene and confirmed successful knockdown of up to 80% at both the transcript and protein levels, using qRT-PCR and ELISA analyses. However, this did not cause a significant rise in JH concentration in the haemolymph. This might suggest that there is more than one isoform of the protein in *Blattella*, or that it only requires a minimal background level of the allatostatin to maintain a normal titer of JH. In contrast, RNAi targeting allatostatin-A in the cricket *Gryllus bimaculatus* (Meyering-Vos *et al.*, 2006) and the allostatin receptor in *Diploptera* itself (Lungchukiet *et al.*, 2008) had a much more noticeable effect, with JH levels in the haemolymph rising soon after RNAi treatment in both cases. However, the induced effect of this heightened JH level differed between species; oogenesis and testes development were impeded in *Gryllus*, while there was only a minimal effect on oocyte length noticeable in *Diploptera*. As a member of the only cockroach family known to bear live young (Blaberidae), future RNAi-based experiments in this species might help elucidate the transition towards viviparity.

Blattella, however, is oviparous, and this strategy depends on the formation of several complex structures within the eggshell that combine protection of the yolk and embryo with their need to respire during development. The problem of maintaining sufficient gas exchange for respiration is exacerbated in this species because it oviposits into an egg case (ootheca), and this has led to the evolution of a structure known as the 'sponge-like body', containing the respiratory aeropyle network (as well as the sperm-entry micropyles) found in most other insect eggshells. The sponge-like body extends from the anterior pole of the egg across the cuticle of the ootheca, forming a line of aeropyles across the keel of the egg case. Irles and co-workers developed an ovary-specific cDNA library (Irles *et al.*, 2009a) to investigate post-vitellogenesis and identified an abundant transcript that was not significantly similar to any known sequences (Irles *et al.*, 2009b). Expressed only during late choriogenesis in the developing oocyte, RNAi-mediated knockdown of the transcript in mature females resulted in defective sponge-like bodies on their eggs. SEM images revealed that the aeropyle network had mostly collapsed, and the authors reasoned that their novel transcript most likely encodes an essential structural protein of the sponge-like body matrix. None of the eggs were viable, for two reasons: collapse of the associated micropyle ensured that none of the eggs could be fertilized and, in any case, they could not have subsequently achieved sufficient gas exchange for development. The authors also noted that the oothecae of knockdown mothers were initially lighter in colour than those laid by control insects. They attributed this to a slower tanning process resulting from the restricted gas exchange and named the novel transcript *brownie*.

RNAi has also recently been applied to the study of speciation via reproductive isolation. The *Allonemobius socius* complex of crickets contains several species and is a model system in the study of ecology and evolutionary biology. These cricket species remain isolated due to postmating, prezygotic factors (including the ability of males to induce female egg laying). Marshall *et al.* (2009) exploited the fact that males of these species package their sperm and accessory gland fluid into a spermatophore, the duct of which they then thread into the female reproductive tract. Spermatophore contents constitute a complete ejaculate, as they have already left the body of the male, but have yet to be mixed with female-specific proteins along her reproductive tract. Marshall and co-workers isolated a highly abundant protein from this uncontaminated ejaculate and subsequently linked it to male reproductive senescence. Older males exhibited less reproductive success than their younger competitors and their ejaculate also contained smaller amounts of the protein of interest. RNAi silencing of the transcript in mature males proved causality, with the females mated with knockdown males laying fewer eggs per day than females that were mated with controls. The transcript was found to encode a serine protease-like enzyme, which the authors dubbed EJAC-SP, but its function within the female reproductive tract is still unknown. It seems likely that it cleaves a specific protein, which in turn initiates the egg-laying process.

4 Neurobiology and behaviour

The neurological control of behaviour is among the most exciting topics in modern biology and much research in this fields has traditionally utilized a small number of well-known hemimetabolous insect models. The grasshoppers and crickets (particularly *Schistocerca spp.* and *Gryllus bimaculatus*) and the American cockroach, *Periplaneta americana*, are popular because of their large size, yielding substantial neurons that lend themselves to the manipulative approaches required for neurophysiology (Burrows, 1996).

Marie *et al.* (2000) focused on developmental neuroplasticity in *Periplaneta* and, in doing so, became the first group to publish results of an RNAi-led experiment in any hemimetabolous insect. *Periplaneta* emerges from the egg with two posterior cerci, each of which bears two wind-sensitive hairs in first instars. Deflection of these hairs transmits a signal to the terminal abdominal ganglion of the CNS, via sensory neurons. During the first moult, the number of hairs increases to 39, with the additional associated sensory neurons forming between the cerci and terminal ganglion. The group had previously ascertained that sensory neurons on the medial side of the cerci express *engrailed*, whereas those on the lateral side do not (Marie and Bacon, 2000), and they hypothesized that this transcription factor must therefore play a role in neuronal differentiation. RNAi silencing of *engrailed*, via injection into the cerci, proved this to be the case. In knockdown animals, axonal projection and arboration of the medial neuron was indistinguishable from that of its lateral counterpart, with key characteristics being the trajectory of the main axon itself and the degree of dorsal branching. *engrailed* remains the only known factor controlling the differentiation of medial sensory cercal neurons from a default, lateral neuron-like phenotype. The group followed this by using RNAi to demonstrate that *engrailed* is not necessary for neuronal specification during embryogenesis, but plays a consistent role after hatching (Marie *et al.*, 2002).

This work was taken a stage further with investigation of the behavioural implications of this neuronal miswiring. Initially, Domenici *et al.* (2008) characterized anti-predator escape responses in *Periplaneta*. American cockroaches detect predators through the perception of sudden air movements by their cercal hairs (Burrows, 1996), and Domenici and co-workers mimicked these turbulences mechanically. Cockroaches would tend to 'escape' from a wind stimulus in an arena by running along one of four seemingly pre-programmed trajectories, at approximately 90°, 120°, 150° and 180° directions away from the perceived predator. Individuals did not significantly prefer any one route over another; they 'kept predators guessing'. Control individuals stimulated from a posterior direction generally chose trajectories that increased the angle between stimulus and escape towards 180°; in other words, they turned away from the perceived predator. This was not the case with the *engrailed*-knockdown animals, which tended to opt for trajectories that decreased the angle between stimulus and escape (Booth *et al.*, 2009). These were not necessarily maladaptive 'wrong moves', as they did not turn *into* the wind (i.e. run towards the predator) any more than the control animals. Nevertheless, the authors show that the responses of these knockdown animals with respect to a posterior stimulus was more in line with the anterior wind responses in the control animals. This was the first paper to link a single transcription factor, which controls synaptic connectivity, to the perception of a directional stimulus.

Interestingly, Marie *et al.* (2000) made the observation that engrailed staining was not affected within the CNS after dsRNA injection into the cerci and suggested that dsRNA could not effectively enter the brain across the perineurium and glia of the blood–brain barrier. This concern has been raised for other animals, including vertebrates (e.g. Thakker *et al.*, 2006) but the blood–brain barrier has since been found to be penetrable to systemic RNAi in a number of insect species, including the aforementioned hemimetabolous models of neurobiology: cockroaches, grasshoppers and crickets.

Among the first to do so were Meyering-Vos and Müller (2007), who targeted sulfakinins with RNAi in the cricket *Gryllus bimaculatus*. The role of these neuropeptides as satiety promoters was first demonstrated in experiments employing the locust *Schistocerca gregaria* and the cockroach *Blatella germanica* (Maestro *et al.*,

2001; Wei *et al.*, 2000), which both ate less when sulfakinin titer in the haemolymph was increased experimentally. Meyering-Vos and Müller introduced dsRNA into *Gryllus* via abdominal injection, or by simply adding it to the animals' food, and used RT-qPCR to confirm that both techniques induced a measurable knockdown in sulfakinin transcript levels in the brain. Female crickets subjected to either mode of RNAi treatment ate significantly more than controls, confirming the satiety-promoting role of this protein in a further species of insect. Perhaps the most remarkable aspect of this experiment, however, was the authors' success in silencing CNS-based transcripts via feeding-induced RNAi. This suggests that the dsRNA persisted long enough in the degrading environment of the digestive tract to be absorbed into the haemolymph and, from there, continued on to be transported across the blood–brain barrier. This appears to be the first such record of ingested dsRNA entering the CNS in any insect (Huvenne and Smagghe, 2010) and has profound implications for the future of neuroethological research.

RNAi studies targeting the hemimetabolous CNS have also investigated some of the more general processes involved in neuronal signalling, including the synthesis and degradation of neurotransmitters. Revuelta *et al.* (2009) targeted the enzyme acetylcholinesterase (AChE) within the thoracic ganglia of *Blattella germanica*, in a pesticide-resistance study. AChE operates at the synaptic cleft, where it terminates neural transmission by hydrolyzing acetylcholine. Organophosphate pesticides irreversibly block AChE, ensuring that neurons cannot repolarize; this eventually kills the affected insect. However, mutations in AChE which lead to less efficient binding by the organophosphate are strongly selected for and populations eventually evolve resistance to the treatment. Furthermore, previous work in *Blattella* found that AChE is actually encoded by two paralogous genes in this species; *ace1* and *ace2*. Topical treatment with insecticide after RNAi-mediated knockdown of either *ace1* or *ace2* in the thoracic ganglia caused a significant difference in mortality, with *ace1*-knockdown animals dying faster. Surprisingly, a*ce2*-knockdown induced mortality was not significantly different from the death rate among control cockroaches. This research demonstrates a possible functional difference between the two paralogues of AChE and suggests that organophosphate resistance might only be due to mutations in *ace1* in this species. The same pattern was noted in *Locusta migratoria*, in a similar experiment by Zhou and Xia (2009). In this species, RNAi against *ace1* via abdominal injection increased sensitivity to an organophosphate insecticide (Malathion) but whether the effect was mediated via reduced expression in the CNS, as was the case mentioned above in *Blatella* (Revuelta *et al.*, 2009), is unclear. It is possible that the fatally interactive effect of *ace1*-RNAi and Malathion in *Locusta* was instead mediated by reduced *ace1* expression in the peripheral nervous system. In any case, both studies were able to attribute insecticide resistance to mutations in *ace1*, which is in contrast to many other insects where mutations in *ace2* are thought to be responsible (Zhou and Xia, 2009).

More recent work by Ma *et al.* (2011) targeted the dopamine synthesis pathway in *Locusta migratoria*, in an attempt to unravel the importance of this neurotransmitter in locust swarming behaviour. In this pathway, the enzyme phenylalanine hydroxylase converts phenylalanine to tyrosine, which is then converted by tyrosine hydroxylase to L-DOPA, the immediate precursor to dopamine. These enzymes are encoded by *henna* and *pale*, respectively, and RNAi against these sequences in *Locusta* reduced the level of swarming-related gregarious behaviour in third-instar nymphs, as measured in a standardized behavioural assay (see Roessingh *et al.*, 1993). However, it is unclear whether the RNAi actually penetrated and affected the CNS, since RT-qPCR was only performed on whole-head tissues and not solely the brain. It seems possible that dopamine levels were only knocked down in the periphery, and so while this would have reduced the locusts' activity levels (and thus their perceived 'phase') it does not necessarily implicate dopamine as the neurotransmitter mediating the underlying neuronal restructuring of the CNS that is expected during behavioural phase change (Ott and Rogers, 2010; Pener and Simpson, 2009) in this species of locust.

A more widely studied process concerned with neuronal restructuring and development is that of learning and formation of both long- and short-term memory. Nitric oxide (NO) plays a role in the development of long-term memory, by directing synapse growth via the well-characterized protein kinase A and CREB-mediated pathway. NO is produced by the enzyme nitric oxide synthase, NOS, and Takahashi *et al.* (2009) used *in situ* hybridization in *Gryllus* to determine that NOS is highly expressed in a subset of Kenyon cells in the mushroom body, a region of the cricket brain involved in olfactory learning and memory. The authors then devised a simple olfactory conditioning test, in which dehydrated crickets were repeatedly subjected to a strong scent of peppermint oil upon access to water. After further water-deprivation, these animals would later opt to explore peppermint-associated holes in a behavioural arena instead of vanilla-scented ones and they retained this association for over 24h. Crickets subject to NOS interference during the conditioning period, via either pharmacological agents or *NOS*-RNAi, also succeeded in selecting the 'correct' hole after 30min, but failed when assayed the following day. This indicates that their ability to form short-term olfactory associations had not been impaired, while the transcription-dependent remodelling of specific synapses, necessary for long-term memory formation, had been inhibited by *NOS* silencing.

5 Circadian rhythms

The oscillatory mechanism of auto-inhibitory feedback loops that constitutes the circadian clock has been studied in *Drosophila* for many years and is well characterized at the molecular level in this species (Tomioka *et al.*, 2009). In essence, the *Drosophila* clock is very simple. During the latter part of the (perceived) daytime, accumulated protein products of *Clock* and *cycle* dimerize in certain cells of the cerebral lobes, enter the nuclei, and up-regulate the transcription of *period* and *timeless*. In turn, the translated products of these genes dimerize towards the latter part of the night and, after complexing with the protein Double-Time, suppress their own transcription by inhibiting *Clock* and *cycle*. The cycle resets once levels of *period/timeless* have diminished, allowing Clock/cycle dimers to accumulate in the cytoplasm once again. Two other feedback loops help to fine-tune the process (Kamae *et al.*, 2010; Tomioka *et al.*, 2009).

However, prior to RNAi little was known about circadian rhythms in other species of insect. In order to compare the process across different insect taxa, Tomioka and co-workers have started to unravel the circadian clock in the cricket *Gryllus bimaculatus*. Moriyama *et al.* (2008, 2009) used RT-qPCR to show that the *period* homologue is expressed rhythmically in *Gryllus* adults and nymphs, before silencing the transcript via abdominal dsRNA injection. Locomotor activity was then measured for up to 40 days on an actograph. While the animals maintained their circadian locomotor activity when maintained under a light-dark cycle, they became arrhythmic when kept solely in the dark. This is in contrast to control insects, which maintained their circadian locomotor activity irrespective of the lighting regime. It would therefore seem that *period* has a conserved role in circadian rhythm generation in *Gryllus* and *Drosophila*. Interestingly, the *Drosophila*-like transcriptional profile of *period* was the same between adult and nymphal crickets, peaking during the subjective night. This is in spite of the fact that nymphs are nocturnal, while the adults tend to be most active during the day (Moriyama *et al.*, 2009).

Despite the apparently conserved role of *period* within the circadian clock, its role is divergent for the related mechanism of diapause. The cricket *Modicogryllus siamensis* expresses a strong photoperiodic regulation of nymphal development, where animals exposed to longer subjective days develop more quickly than those reared with a longer nighttime. Diapause is a common strategy among temperate zone insects, and ensures that individuals do not tend to reach reproductive maturity while conditions are unfavourable. Sakamoto *et al.* (2009) showed that nymphs of *Modicogryllus* reared under a 16-h light, 8-h dark regime (LD16:8) took an average of 50 days to reach maturity, while reversal to a LD8:16 program extended this duration to around 140 days. *Modicogryllus* nymphs reared entirely in the dark showed a much wider range of phenotypes, with individuals taking 40–160 days to reach adulthood. Injecting *period* dsRNA into adult females created a curious phenotype in their offspring which, independent of light-dark rearing regime, developed to adulthood in a similar pattern to the light-deprived animals. This suggests that the circadian rhythm machinery in this cricket is a prerequisite for photoperiodic time measurement, with *period* playing a central role. This appears to be different from *Drosophila*, in which *period* mutants still exhibit photoperiodicity in line with the wild type (Sakamoto *et al.*, 2009; Saunders, 1990; Tomioka *et al.*, 2009).

Characterization of the *Gryllus* homologue of *timeless* has revealed yet further divergence of function within the circadian machinery (Danbara *et al.*, 2010). Initial RT-PCR in adults and nymphs showed that *timeless* expression mimics that of *period*, peaking during the dark cycle as it does in *Drosophila*, and *timeless*-RNAi was successful in lowering transcript levels in the optic lobes to the point of arrhythmicity. Surprisingly, however, knockdown crickets maintained their circadian rhythm in spite of this silencing, though their level of activity during the subjective daytime period (or nighttime in the case of nymphs) was lower than that exhibited by control animals. *Gryllus timeless* therefore appears to play some role in the fine-tuning of circadian locomotor activities, but is not essential for maintaining the rhythmic oscillations of the circadian clock in this cricket.

To help identify the ancestral function for clock genes, Tomioka's group has also started to investigate the endogenous clock in the apterygote firebrat *Thermobia domestica*, which is equally distantly related to all the winged insects (Kamae *et al.*, 2010). RNAi against the genes *Clock* and *cycle*, by direct injection into the abdomen of adult insects, abolished rhythmic cycles for up to 30 days in treated individuals kept solely in the dark. However, a LD12:12 regime was sufficient to maintain a circadian rhythm in *cycle*-knockdown individuals, while *Clock*-knockdown continued to induce complete arrhythmicity. The authors postulate that this pattern places the *Clock* gene product in the light input pathway in this insect, though the actual circadian centre is still to be determined. The function of *Clock* also remains elusive in the desert locust *Schistocerca gregaria* and was investigated with RNAi by Tobback *et al.* (2011). Knockdown of *Clock* proved fatal for all males, which were injected as part of a study into the importance of clock genes in reproductive capacity. By contrast, RNAi against *period* and *timeless* in females resulted in their producing fewer progeny than controls.

Circadian rhythmicity has also been investigated in *Blattella*. Lee *et al.* (2009) investigated the role of pigment-dispersing factor (PDF), a neuropeptide that is synthesized constantly in the circadian centre of the optic lobes. Silencing of the *pdf* transcript lasted for more than 50 days after a single abdominal injection of dsRNA and caused around half of the knockdown cockroaches to become arrhythmic under a LD12:12 regime, while around 90% of them exhibited circadian impairment under constant darkness. This clearly demonstrates a role of *pdf* in *Blattella* circadian rhythm generation, which again differs from that in *Drosophila*; in the fruit fly, *pdf* mutants only lose their circadian locomotor rhythm under constant darkness.

Continued RNAi-mediated investigation of the hemimetabolous (and ametabolous) circadian clock seems a particularly important avenue for future research, given the apparent variation from the *Drosophila*-biased paradigm that has resulted from the work done so far. Results of this nature highlight the necessity of cross-taxa comparisons and the potential of RNAi in investigations of this kind. Tomioka's work has elegantly highlighted the utility of *Gryllus bimaculatus* as a workhorse of RNAi-driven experiments; in some cases, the effects of a single injection of dsRNA lasted for more than 2 months (Tomioka *et al.*, 2009). Nevertheless, the extent to which dsRNA penetrates the blood–brain barrier remains uncertain (e.g. Marie *et al.*, 2000) and future work could also seek to develop ways of reliably introducing RNAi across the neural sheath, possibly through microsurgery, electroporation or via a lipid-carrier molecule (discussed for vertebrate therapeutics in Jagannath and Wood, 2007).

6 Feeding

RNAi was first described in animals in the nematode *C. elegans*, and it was quickly ascertained that its effects could be induced via feeding of the dsRNA (Fire *et al.*, 1998; Timmons and Fire, 1998; Timmons *et al.*, 2001). This was successfully reported for the insects some time later, initially in the lepidopteran light brown apple moth *Epiphyas postvittana* and the hemipteran triatomine bug *Rhodnius prolixus* (Araujo *et al.*, 2006; Turner *et al.*, 2006). The marriage of insect nutritional science and RNAi has since explored two broad avenues of investigation. Firstly, the process of digestion and nutrient regulation itself has been investigated in its own right, with dsRNA being delivered through either feeding or more conventional means (e.g. microinjection). Secondly, RNAi-induced lethality has led to the technique being mooted as a novel means of insect pest control. As one might expect, there is considerable overlap between the latter sphere of research and all other RNAi-utilizing studies: any gene found to increase mortality when silenced, either by accident or design, is lauded as a potential defender of crops. The mechanisms through which this might be possible have been investigated and reviewed in great detail elsewhere (see Baum *et al.*, 2007; Gordon and Waterhouse, 2007; Huvenne and Smagghe, 2010; Mito *et al.*, 2011; Price and Gatehouse, 2008; Whyard *et al.*, 2009). This section will concentrate on studies that have used RNAi to decipher some part of the feeding or digestion process.

The first successful demonstration of feeding-mediated RNAi in the hemimetabolous insects was in the triatomine bug *Rhodnius prolixus* (Araujo *et al.*, 2006). Along with other closely related species in the family Reduviidae (Hemiptera), these insects are the most important vectors of Chagas disease (or American trypanosomiasis) in Central and South America. Exclusively haematophagous, they are known colloquially as 'kissing bugs' after their propensity to take human blood meals from the soft tissue of the lips, usually when the host is asleep. It is during this process that the protozoan pathogen, *Trypanosoma cruzi*, can transfer from an infected human to the insect vector, while transmission in the other direction occurs through human contact with the insects' faecal deposits (Speight *et al.*, 2008). While initial symptoms of Chagas disease are generally mild, they can progress towards complications of the heart and digestive system if left untreated. Unfortunately this is all too often the case, with an estimated 8–10 million people in developing, rural parts of the Americas unaware that they have the illness (Bern *et al.*, 2007). Not surprisingly, assassin bugs are well studied across this region. Araujo *et al.* (2006) introduced dsRNA via both injection and feeding to down-regulate expression of *nitrophorin 2* in the salivary gland. The nitrophorins store and transport nitric oxide that, upon injection into the circulatory system of a mammalian host, causes anticoagulant effects including reduced platelet aggregation and vasodilation. Knockdown of *nitrophorin 2* caused a reduction of the protein in the saliva, leading to a severe reduction in anticoagulant ability (control insects could prolong the process for around four times longer than the treatment group). This work was built upon by electronically recording *Rhodnius* behaviour while feeding on a host (Araujo *et al.*, 2009). *Nitrophorin*-knockdown animals bit around five times more often than control insects, yet still had a lower cumulative 'feeding time' in which they were settled and successfully imbibing a blood meal. Nitrophorins are therefore essential for efficient feeding activity, in which minimal bites are required to decrease the likelihood of being noticed by the host.

Beyond the action of the salivary glands, blood meals taken by the assassin bugs need to be further processed along the digestive tract. Turning their attention to a related vector of Chagas disease, Pereira's group identified a novel thrombin inhibitor in the intestine of *Triatoma brasiliensis* and implicated it in blood intake and processing (Araujo *et al.*, 2007). Thrombin promotes blood coagulation and RNAi knockdown of the novel thrombin inhibitor (dubbed *brasiliensin*) led to affected nymphs ingesting around 40% less blood than the control group. Despite this pattern, blood meal clotting in the midgut did not seem to affect digestion and knockdown insects still survived, moulted and fed again given the opportunity. The authors speculate that the reduced blood intake seen in the *brasiliensin*-RNAi insects is perhaps due to a coagulation-induced increase in viscosity of the intestinal contents, leading to a reduced efficiency of buccal pumping through the stylet. This research highlights the importance of a sustained anticoagulation mechanism throughout the entire digestive system in haematophagous insects and not just at the puncture wound of the host.

The blood-sucking assassin bugs notwithstanding, the majority of the hemipteran order feed on the phloem of plants and often with major consequences for agriculture. The pea aphid *Acyrthosiphon pisum* is a pest of various legumes, though it is not as economically important as some closely related species (Godfray, 2010). Nevertheless, it has recently become the first hemimetabolous insect to have its genome fully sequenced and annotated, paving the way for multiple comparative studies with the previously sequenced insect species that undergo a full metamorphosis (International Aphid Genomics Consortium, 2010). *Acyrthosiphon* has already proven amenable to gene silencing with RNAi. Injection of dsRNA led to a 40% decrease in gene expression for the housekeeping gene *calreticulin* and the gut-specific protease *cathepsin*, though no phenotypic change was apparent in either case (Jaubert-Possamai *et al.*, 2007). Incidentally,

Liu *et al.* (2010) had similar results in the brown planthopper *Nilaparvata lugens*, a major hemipteran rice pest in many parts of Asia. They also induced a 40% reduction in *calreticulin* and *cathepsin* expression, as well as a 25% knockdown of the CNS gene N1β2, also without documented phenotypic effect in any case.

Mutti *et al.* (2006) were more successful in evoking a measurable RNAi-led phenotypic change in *Acyrthosiphon* by down-regulating the protein C002, which had previously been found in abundance in the salivary gland. RNAi against its transcript led to a significant increase in aphid mortality, which was attributed to an inability to feed on the host plant, fava bean (Mutti *et al.*, 2008). This experiment used the electrical penetration graph (EPG) method, in which a gold electrode is affixed to the dorsum of the aphid with conductive adhesive, and a larger electrode is inserted into the soil of the host plant. Contact between the insect stylet and plant closes a simple circuit, yielding an interpretable electrical signal indicating the number and duration of stylet insertions. C002-knockdown aphids were drastically unsuccessful feeders; they probed the plant cells, but did not locate the phloem in most cases and, even if they did, they did not maintain penetration of the stylet. Mutti and co-workers also used *in situ* hybridization and immunohistochemistry to locate the expression of C002 protein to within a specialized subset of cells in the salivary gland, and also in the host plant leaf after successful pea aphid feeding. The crucial mechanism through which this salivary gland protein operates is yet to be determined (Mutti *et al.*, 2008).

Aphid feeding and osmoregulation are necessarily coordinated tightly, as all nutrients are imbibed in solution through the stylet. Plant phloem sap usually has a higher osmotic potential than the haemolymph of aphids that feed on them, and therefore has the potential to dehydrate the body of the insect by drawing water into the gut lumen during stylet puncture (Shakesby *et al.*, 2009). This issue can be mitigated by water cycling in the simple aphid gut, in which water from the distal end of the intestine can diffuse back into the stomach. Shakesby and co-workers identified and characterized an aquaporin protein, *AQP1*, implicated in the structure of selectively permeable water channels that regulate this aphid water cycling. The group introduced *AQP1* dsRNA through feeding (the first time it has been shown in the pea aphid), since fluid injection, as well as the associated cuticle perforation, could have had adverse effects on osmotic pressures in the insect. Knockdown animals exhibited a transient yet significant elevation of osmotic pressure in the haemolymph by up to 28%, proving that this protein is involved in homeostatic water cycling in this insect.

Beyond the sucking bugs, very little RNAi work appears to have targeted feeding-related transcripts in the hemimetabolous insects. In the termite *Reticulitermes*, Zhou *et al.* (2008) focused on digestion by targeting the cellulase enzyme, Cell-1. In a simple feeding bioassay, groups of 15 worker termites were given access to discs of paper towel soaked with *Cell*-1-dsRNA. Treated insects tended to eat more of the paper discs, which the authors interpreted as a compensatory mechanism due to a lack of cellulose digestion. In any case, this heightened level of feeding activity did not overcome the inability to digest cellulose, with treated colonies exhibiting a significant increase in mortality after as little as 16 days. Feeding-mediated *Hex-2* silencing also led to a significant rise in mortality, possibly due to disruption of the social structure that arose from an increase in worker-soldier development. With worldwide termite damage to human structures estimated at around US$20 billion per annum, it seems likely that novel target sequences for future termiticides will arise from further feeding trials of this nature.

The impact of insects on people has also been the focus of work by Schal's group, who are particularly concerned with the behaviour and physiology of pest cockroach species. The model cockroach *Blattella germanica* is one such animal, which produces various allergens in its faeces that can cause or exacerbate human respiratory conditions, including asthma. Suazo *et al.* (2009) used RNAi to down-regulate one such allergen, the protein Bla g 1, to characterize its functions within *Blattella*. Knockdown insects exhibiting a slower rate of midgut growth, which in turn resulted in reduced weight gain over the treatment period. This digestive disruption also impeded oocyte growth and development, though this was rescued by treating affected females with JH. The importance of this single protein in *Blattella* growth also has clear implications for the development of novel control strategies.

7 Future research and concluding remarks

RNAi has yielded some fascinating insights into the biology of those insects without a genome sequence, but the wealth of opportunity yet to be explored is surely more exciting still. Approximately 75% of all named insects species occur within the nine holometabolous insect orders, with the remaining numbers spread across 18 orders that do not undergo full metamorphosis (Mito *et al.*, 2010). Only five of these 18 hemimetabolous or ametabolous orders have been investigated with RNAi (Fig. 1), with the remaining groups accounting for some interesting and economically important insect species. These include the stoneflies (Plecoptera), the only insects in which the haemolymph contains both hexamerin storage proteins, and the respiratory hemocyanins from which they are thought to have evolved (Hagner-Holler *et al.*, 2007). Hexamerins have already been discussed with regard to their role in termite caste differentiation, and RNAi in the stoneflies could reveal yet more about the origins and functions of these proteins. Praying mantises (Mantodea) are the sister order to the cockroaches and termites, two orders which, as this review has shown, appear to be particularly amenable to RNAi. As a notably carnivorous taxon within the hemimetabolous insects, RNAi could contribute towards the neuroethology of mantid predatory behaviour. The biology of extremophiles could be investigated using RNAi in the ice crawlers (Grylloblattodea), which exist in some of the coldest regions of the Northern Hemisphere (McGavin and Lewington,

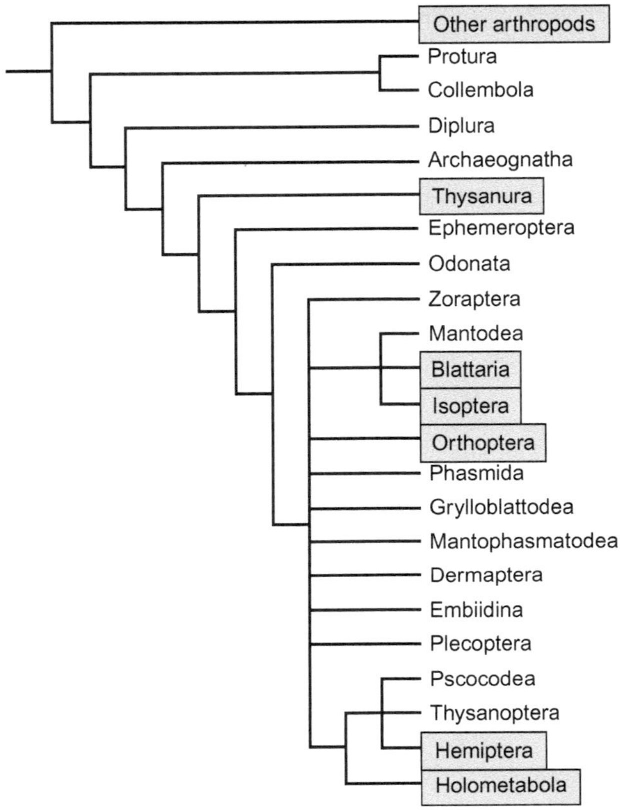

FIG. 1 A simple phylogeny of the known orders of hexapod. The insects that undergo a full metamorphosis have been collapsed into a single taxon, 'Holometabola', for simplicity. Orders in which RNA interference has been successfully performed are highlighted.

2001). The web-spinners (Embiidina) are also surely worthy of investigation; their silk glands providing a structure with which functional genomic comparisons can be made to spiders and the lepidopteran silkworms, including the model species for sericulture *Bombyx mori* (Sutherland *et al.*, 2010). There are also some conspicuous omissions towards the taxonomic root of the insect phylogeny, where evolutionary relationships between the extant orders of insects, and the other classes of arthropod, are the topic of much ongoing debate (Grimaldi, 2010; Regier *et al.*, 2010). Successful RNAi-based studies have yet to be reported for work on dragonflies (Odonata), mayflies (Ephemeroptera), or bristletails (Archaeognatha), and nor has it been recorded for any of the related hexapod outgroups (the entognathan Diplura, Collembola and Protura). As this review has highlighted, RNAi has revealed myriad genes that have evolved divergent functions between taxa, and similar discoveries must surely await our attention within the outgrouping classes of arthropod.

Perhaps most surprising, however, are the neglected orders that include some of the most economically important insects. Many species of booklice (Psocoptera) and thrips (Thysanoptera) are serious pests of crop plants and stored products (McGavin and Lewington, 2001). Recent EST analysis of the western flower thrip, *Frankliniella occidentalis*, has revealed the presence of multiple proteins necessary in the RNAi pathway (Rotenberg and Whitfield, 2010), but studies employing the technique to elucidate gene function in these animals have yet to be published. Also worthy of mention in the context of negative human interaction is the body louse, *Pediculus humanus humanus*, which feeds solely on human blood and is an important vector species for diseases including typhus and trench fever. The genome of this insect was recently published, alongside the genome of its obligate bacterial symbiont (provisionally named *Riesia pediculicola*), from which the louse derives nutrients to supplement the human blood meal (International Louse Genomics Consortium, 2010). Initial analyses of the louse genome place it as the smallest known in any insect (108 Mb), having lost many of the environmental sensing genes associated with a free-living life history. RNAi has yet to be demonstrated in *Pediculus* but, alongside the recently genome-sequenced pea aphid *Acyrthosiphon pisum* (International Aphid Genomics Consortium, 2010), greater functional genomic comparison between the hemimetabolous and holometabolous orders through RNAi is surely imminent. Furthermore, the ever-increasing popularity of RNAi might yet establish a non-metamorphosing insect species as a rival model to *Drosophila* for functional genomics, given the relatively low amenity of the fruit fly towards dsRNA knockdown via simple injection (Bellés, 2010). The most exciting discoveries look set to arrive in the very near future.

Acknowledgements

I would like to thank Professor Stephen J. Simpson, Professor Jérôme Casas, and Associate Professor Elizabeth Jockusch for the invitation to contribute to this volume. Dr. Nathan Lo and two anonymous reviewers read and commented on a previous draft of the manuscript, and I am very grateful for their efforts. D.A.C. was funded by a University of Sydney International Postgraduate Research Scholarship.

References

Anderson, H. and French, V. (1985). Cell division during intercalary regeneration in the cockroach leg. *J. Embryol. Exp. Morphol.* **90**, 57–78.

Angelini, D. R. and Kaufman, T. C. (2004). Functional analyses in the hemipteran *Oncopeltus fasciatus* reveal conserved and derived aspects of appendage patterning in insects. *Dev. Biol.* **271**, 306–321.

Angelini, D. R. and Kaufman, T. C. (2005a). Functional analyses in the milkweed bug *Oncopeltus fasciatus* (Hemiptera) support a role for Wnt signaling in body segmentation but not appendage development. *Dev. Biol.* **283**, 409–423.

Angelini, D. R. and Kaufman, T. C. (2005b). Insect appendages and comparative ontogenetics. *Dev. Biol.* **286**, 57–77.

Angelini, D. R., Liu, P. Z., Hughes, C. L. and Kaufman, T. C. (2005). Hox gene function and interaction in the milkweed bug *Oncopeltus fasciatus* (Hemiptera). *Dev. Biol.* **287**, 440–455.

Araujo, R. N., Santos, A., Pinto, F. S., Gontijo, N. F., Lehane, M. J. and Pereira, M. H. (2006). RNA interference of the salivary gland nitrophorin 2 in the triatomine bug *Rhodnius prolixus* (Hemiptera: Reduviidae) by dsRNA ingestion or injection. *Insect Biochem. Mol. Biol.* **36**, 683–693.

Araujo, R. N., Campos, I. T. N., Tanaka, A. S., Santos, A., Gontijo, N. F., Lehane, M. J. and Pereira, M. H. (2007). Brasiliensin: a novel intestinal thrombin inhibitor from *Triatoma brasiliensis* (Hemiptera: Reduviidae) with an important role in blood intake. *Int. J. Parasitol.* **37**, 1351–1358.

Araujo, R. N., Soares, A. C., Paim, R. M. M., Gontijo, N. F., Gontijo, A. F., Lehane, M. J. and Pereira, M. H. (2009). The role of salivary nitrophorins in the ingestion of blood by the triatomine bug *Rhodnius prolixus* (Reduviidae: Triatominae). *Insect Biochem. Mol. Biol.* **39**, 83–89.

Bando, T., Mito, T., Maeda, Y., Nakamura, T., Ito, F., Watanabe, T., Ohuchi, H. and Noji, S. (2009). Regulation of leg size and shape by the Dachsous/Fat signalling pathway during regeneration. *Development* **136**, 2235–2245.

Baum, J. A., Bogaert, T., Clinton, W., Heck, G. R., Feldmann, P., Ilagan, O., Johnson, S., Plaetinck, G., Munyikwa, T., Pleau, M., Vaughn, T. and Roberts, J. (2007). Control of coleopteran insect pests through RNA interference. *Nat. Biotechnol.* **25**, 1322–1326.

Bellés, X. (2010). Beyond Drosophila: RNAi in vivo and functional genomics in insects. *Annu. Rev. Entomol.* **55**, 111–128.

Ben-David, J. and Chipman, A. D. (2010). Mutual regulatory interactions of the trunk gap genes during blastoderm patterning in the hemipteran *Oncopeltus fasciatus*. *Dev. Biol.* **346**, 140–149.

Bern, C., Montgomery, S. P., Herwaldt, B. L., Rassi, A. Jr., Marin-Neto, J. A., Dantas, R. O., Maguire, J. H., Acquatella, H., Morillo, C., Kirchhoff, L. V., Gilman, R. H., Reyes, P. A., *et al.* (2007). Evaluation and treatment of Chagas disease in the United States: a systematic review. *J. Am. Med. Assoc.* **298**, 2171–2181.

Bolognesi, R., Beerman, A., Farzana, L., Wittkopp, N., Lutz, R., Balavoine, G., Brown, S. J. and Schroder, R. (2008). Tribolium Wnts: evidence for a larger repertoire in insects with overlapping expression patterns that suggest multiple redundant functions in embryogenesis. *Dev. Genes Evol.* **218**, 193–202.

Booth, D., Marie, B., Domenici, P., Blagburn, J. M. and Bacon, J. P. (2009). Transcriptional control of behavior: engrailed knock-out changes cockroach escape trajectories. *J. Neurosci.* **29**, 7181–7190.

Braun, R. P. and Wyatt, G. R. (1996). Sequence of the hexameric juvenile hormone-binding protein from the hemolymph of *Locusta migratoria*. *J. Biol. Chem.* **271**, 31756–31762.

Bucher, G. and Klingler, M. (2004). Divergent segmentation mechanism in the short germ insect *Tribolium* revealed by giant expression and function. *Development* **131**, 1729–1740.

Burmester, T. (1999). Evolution and function of the insect hexamerins. *Eur. J. Entomol.* **96**, 213–225.

Burrows, M. (1996). The Neurobiology of an Insect Brain. Oxford University Press, Oxford, UK.

Caygill, E. E. and Johnston, L. A. (2008). Temporal regulation of metamorphic processes in Drosophila by the let-7 and miR-125 heterochronic microRNAs. *Curr. Biol.* **18**, 943–950.

Chesebro, J., Hrycaj, S., Mahfooz, N. and Popadić, A. (2009). Diverging functions of Scr between embryonic and post-embryonic development in a hemimetabolous insect, *Oncopeltus fasciatus*. *Dev. Biol.* **329**, 142–151.

Ciudad, L., Piulachs, M.-D. and Bellés, X. (2006). Systemic RNAi against the cockroach vitellogenin receptor results in a phenotype similar to that of the *Drosophila* yolkless mutant. *FEBS J.* **273**, 325–335.

Ciudad, L., Bellés, X. and Piulachs, M.-D. (2007). Structural and RNAi characterization of the German cockroach lipophorin receptor, and the evolutionary relationships of lipoprotein receptors. *BMC Mol. Biol.* **8**, 53.

Cruz, J., Mané-Padrós, D., Bellés, X. and Martín, D. (2006). Functions of the ecdysone receptor isoform-A in the hemimetabolous insect *Blattella germanica* revealed by systemic RNAi in vivo. *Dev. Biol.* **297**, 158–171.

Cruz, J., Martín, D. and Bellés, X. (2007). Redundant ecdysis regulatory functions of three nuclear receptor HR3 isoforms in the direct-developing insect *Blattella germanica*. *Mech. Dev.* **124**, 180–189.

Cruz, J., Nieva, C., Mané-Padrós, D., Martín, D. and Bellés, X. (2008). Nuclear receptor BgFTZ-F1 regulates molting and the timing of ecdysteroid production during nymphal development in the hemimetabolous insect Blattella germanica. *Dev. Dyn.* **237**, 3179–3191.

Danbara, Y., Sakamoto, T., Uryu, O. and Tomioka, K. (2010). RNA interference of *timeless* gene does not disrupt circadian locomotor rhythms in the cricket *Gryllus bimaculatus*. *J. Insect Physiol.* **56**, 1738–1745.

De Robertis, E. M. (2008). The molecular ancestry of segmentation mechanisms. *Proc. Natl. Acad. Sci. USA* **105**, 16411–16412.

Domenici, P., Booth, D., Blagburn, J. M. and Bacon, J. P. (2008). Cockroaches keep predators guessing by using preferred escape trajectories. *Curr. Biol.* **18**, 1792–1796.

Dong, Y. and Friedrich, M. (2005). Nymphal RNAi: systemic RNAi mediated gene knockdown in juvenile grasshopper. *BMC Biotechnol.* **5**, 25.

Dong, Y. and Friedrich, M. (2010). Enforcing biphasic eye development in a directly developing insect by transient knockdown of single eye selector genes. *J. Exp. Zool. B Mol. Dev. Evol.* **314**, 104–114.

Edwards, J. S. and Reddy, R. (1986). Mechanosensory Appendages and giant interneurons in the firebrat (*Thermobia domestica*, Thysanura): a prototype system for terrestrial predator evasion. *J. Comp. Neurol.* **243**, 535–546.

Erezyilmaz, D. F., Riddiford, L. M. and Truman, J. W. (2006). The pupal specifier *broad* directs progressive morphogenesis in a direct-developing insect. *Proc. Natl. Acad. Sci. USA* **103**, 6925–6930.

Erezyilmaz, D. F., Rynerson, M. R., Truman, J. W. and Riddiford, L. M. (2009a). The role of the pupal determinant *broad* during embryonic development of a direct-developing insect. *Dev. Genes Evol.* **219**, 535–544.

Erezyilmaz, D. F., Kelstrup, H. C. and Riddiford, L. M. (2009b). The nuclear receptor E75A has a novel pair-rule-like function in patterning the milkweed bug, *Oncopeltus fasciatus*. *Dev. Biol.* **334**, 300–310.

Fire, A., Xu, S. Q., Montgomery, M. K., Kostas, S. A., Driver, S. E. and Mello, C. C. (1998). Potent and specific genetic interference by double-stranded RNA in *Caenorhabditis elegans*. *Nature* **391**, 806–811.

Flockhart, I., Booker, M., Kiger, A., Boutros, M., Armknecht, S., Ramadan, N., Richardson, K., Xu, A., Perrimon, N. and Mathey-Prevot, B. (2006). FlyRNAi: the *Drosophila* RNAi screening center database. *Nucleic Acids Res.* **34**, D489–D494.

Ghanim, M., Kontsedalov, S. and Czosnek, H. (2007). Tissue-specific gene silencing by RNA interference in the whitefly *Bemisia tabaci* (Gennadius). *Insect Biochem. Mol. Biol.* **37**, 732–738.

Godfray, H. C. J. (2010). The pea aphid genome. *Insect Mol. Biol.* **19**, 1–4.

Gomez-Orte, E. and Bellés, X. (2009). MicroRNA-dependent metamorphosis in hemimetabolan insects. *Proc. Natl. Acad. Sci. USA* **106**, 21678–21682.

Gordon, K. H. J. and Waterhouse, P. M. (2007). RNAi for insect-proof plants. *Nat. Biotechnol.* **25**, 1231–1232.

Grimaldi, D. A. (2010). 400 million years on six legs: on the origin and early evolution of Hexapoda. *Arthropod Struct. Dev.* **39**, 191–203.

Hagner-Holler, S., Pick, C., Girgenrath, S., Marden, J. H. and Burmester, T. (2007). Diversity of stonefly hexamerins and implication for the evolution of insect storage proteins. *Insect Biochem. Mol. Biol.* **37**, 1064–1074.

Hamada, A., Miyawaki, K., Honda-Sumi, E., Tomioka, K., Mito, T., Ohuchi, H. and Noji, S. (2009). Loss-of-function analyses of the fragile X-related and dopamine receptor genes by RNA interference in the cricket Gryllus bimaculatus. *Dev. Dyn.* **238**, 2025–2033.

He, Z. B., Cao, Y. Q., Yin, Y. P., Wang, Z. K., Chen, B., Peng, G. X. and Xia, Y. X. (2006). Role of hunchback in segment patterning of *Locusta migratoria manilensis* revealed by parental RNAi. *Dev. Growth Differ.* **48**, 439–445.

He, Z. B., Cao, Y. Q. and Xia, Y. X. (2010). Optimization of parental RNAi conditions for hunchback gene in Locusta migratoria manilensis (Meyen). *Insect Sci.* **17**, 1–6.

Herke, S. W., Serio, N. V. and Rogers, B. T. (2005). Functional analyses of *tiptop* and *Antennapedia* in the embryonic development of *Oncopeltus fasciatus* suggests an evolutionary pathway from ground state to insect legs. *Development* **132**, 27–34.

Hrycaj, S., Mihajlovic, M., Mahfooz, N., Couso, J. P. and Popadić, A. (2008). RNAi analysis of *nubbin* embryonic functions in a hemimetabolous insect, *Oncopeltus fasciatus*. *Evol. Dev.* **10**, 705–716.

Hrycaj, S., Chesebro, J. and Popadić, A. (2010). Functional analysis of *Scr* during embryonic and post-embryonic development in the cockroach, *Periplaneta americana*. *Dev. Biol.* **341**, 324–334.

Hughes, C. L. and Kaufman, T. C. (2000). RNAi analysis of *Deformed*, *proboscipedia* and *Sex combs reduced* in the milkweed bug *Oncopeltus fasciatus*: novel roles for Hox genes in the hemipteran head. *Development* **127**, 3683–3694.

Huvenne, H. and Smagghe, G. (2010). Mechanisms of dsRNA uptake in insects and potential of RNAi for pest control: a review. *J. Insect Physiol.* **56**, 227–235.

International Aphid Genomics Consortium, (2010). Genome sequence of the pea aphid Acyrthosiphon pisum. *PLoS Biol.* **8**, 179–296.

International Louse Genomics Consortium, (2010). Genome sequences of the human body louse and its primary endosymbiont provide insights into the permanent parasitic lifestyle. *Proc. Natl. Acad. Sci. USA* **107**, 12168–12173.

Inward, D., Beccaloni, G. and Eggleton, P. (2007). Death of an order: a comprehensive molecular phylogenetic study confirms that termites are eusocial cockroaches. *Biol. Lett.* **3**, 331–335.

Irles, P., Bellés, X. and Piulachs, M. D. (2009a). Identifying genes related to choriogenesis in insect panoistic ovaries by Suppression Subtractive Hybridization. *BMC Genomics* **10**, 206.

Irles, P., Bellés, X., Piulachs, M. D. and Laudet, V. (2009b). *Brownie*, a gene involved in building complex respiratory devices in insect eggshells. *PLoS One* **4**, e8353.

Jagannath, A. and Wood, M. (2007). RNA interference based gene therapy for neurological disease. *Brief. Funct. Genomic. Proteomic.* **6**, 40–49.

Jaubert-Possamai, S., Le Trionnaire, G., Bonhomme, J., Christophides, G. K., Rispe, C. and Tagu, D. (2007). Gene knockdown by RNAi in the pea aphid *Acyrthosiphon pisum*. *BMC Biotechnol.* **7**, 63.

Jorgensen, R. (1990). Altered gene expression in plants due to trans interactions between homologous genes. *Trends Biotechnol.* **8**, 320–324.
Kamae, Y., Tanaka, F. and Tomioka, K. (2010). Molecular cloning and functional analysis of the clock genes, *Clock* and *cycle*, in the firebrat *Thermobia domestica. J. Insect Physiol.* **56**, 1291–1299.
Kang, L., Chen, X. Y., Zhou, Y., Liu, B. W., Zheng, W., Li, R. Q., Wang, J. and Yu, J. (2004). The analysis of large-scale gene expression correlated to the phase changes of the migratory locust. *Proc. Natl. Acad. Sci. USA* **101**, 17611–17615.
Kennerdell, J. R. and Carthew, R. W. (1998). Use of dsRNA-mediated genetic interference to demonstrate that *frizzled* and *frizzled 2* act in the wingless pathway. *Cell* **95**, 1017–1026.
Khila, A., Abouheif, E. and Rowe, L. (2009). Evolution of a novel appendage ground plan in water striders is driven by changes in the hox gene *Ultrabithorax. PLoS Genet.* **5**, e1000583.
Korb, J. and Hartfelder, K. (2008). Life history and development-a framework for understanding developmental plasticity in lower termites. *Biol. Rev.* **83**, 295–313.
Korb, J., Weil, T., Hoffmann, K., Foster, K. R. and Rehli, M. (2009). A gene necessary for reproductive suppression in termites. *Science* **324**, 758.
Kraut, R. and Levine, M. (1991). Mutually repressive interactions between the gap genes giant and Kruppel define middle body regions of the Drosophila embryo. *Development* **111**, 611–621.
Lee, C.-M., Su, M.-T. and Lee, H.-J. (2009). Pigment dispersing factor: an output regulator of the circadian clock in the german cockroach. *J. Biol. Rhythms* **24**, 35–43.
Liu, P. Z. and Kaufman, T. C. (2004a). *hunchback* is required for suppression of abdominal identity, and for proper germband growth and segmentation in the intermediate germband insect *Oncopeltus fasciatus. Development* **131**, 1515–1527.
Liu, P. Z. and Kaufman, T. C. (2004b). *Krüppel* is a gap gene in the intermediate germband insect *Oncopeltus fasciatus* and is required for development of both blastoderm and germband-derived segments. *Development* **131**, 4567–4579.
Liu, P. Z. and Kaufman, T. C. (2005a). *even-skipped* is not a pair-rule gene but has segmental and gap-like functions in *Oncopeltus fasciatus*, an intermediate germband insect. *Development* **132**, 2081–2092.
Liu, P. Z. and Kaufman, T. C. (2005b). Short and long germ segmentation: unanswered questions in the evolution of a developmental mode. *Evol. Dev.* **7**, 629–646.
Liu, P. Z. and Patel, N. H. (2010). *giant* is a bona fide gap gene in the intermediate germband insect, *Oncopeltus fasciatus. Development* **137**, 835–844.
Liu, S., Ding, Z., Zhang, C., Yang, B. and Liu, Z. (2010). Gene knockdown by intro-thoracic injection of double-stranded RNA in the brown planthopper, *Nilaparvata lugens. Insect Biochem. Mol. Biol.* **40**, 666–671.
Lo, N., Engel, M. S., Cameron, S., Nalepa, C. A., Tokuda, G., Grimaldi, D., Kitade, O., Krishna, K., Klass, K. D., Maekawa, K., Miura, T. and Thompson, G. J. (2007). Save isoptera: a comment on Inward *et al. Biol. Lett.* **3**, 562–563.
Lungchukiet, P., Donly, B. C., Zhang, J., Tobe, S. S. and Bendena, W. G. (2008). Molecular cloning and characterization of an allatostatin-like receptor in the cockroach *Diploptera punctata. Peptides* **29**, 276–285.
Ma, Z., Guo, W., Guo, X., Wang, X. and Kang, L. (2011). Modulation of behavioural phase changes of the migratory locust by the catecholamine metabolic pathway. *Proc. Natl. Acad. Sci. USA* **108**, 3883–3887.
Maestro, J. L. and Bellés, X. (2006). Silencing allatostatin expression using double-stranded RNA targeted to preproallatostatin mRNA in the German cockroach. *Arch. Insect Biochem. Physiol.* **62**, 73–79.
Maestro, J. L., Aguilar, R., Pascual, N., Valero, M. L., Piulachs, M. D., Andreu, D., Navarro, I. and Bellés, X. (2001). Screening of antifeedant activity in brain extracts led to the identification of sulfakinin as a satiety promoter in the German cockroach. *Eur. J. Biochem.* **268**, 5824–5830.

Mahfooz, N. S., Li, H. and Popadić, A. (2004). Differential expression patterns of the hox gene are associated with differential growth of insect hind legs. *Proc. Natl. Acad. Sci. USA* **101**, 4877–4882.

Mahfooz, N., Turchyn, N., Mihajlovic, M., Hrycaj, S. and Popadić, A. (2007). Ubx regulates differential enlargement and diversification of insect hind legs. *PLoS One* **2**(9), e866no. 9.

Mané-Padrós, D., Cruz, J., Vilaplana, L., Pascual, N., Bellés, X. and Martín, D. (2008). The nuclear hormone receptor BgE75 links molting and developmental progression in the direct-developing insect *Blattella germanica*. *Dev. Biol.* **315**, 147–160.

Mané-Padrós, D., Cruz, J., Vilaplana, L., Nieva, C., Ureña, E., Bellés, X. and Martín, D. (2010). The hormonal pathway controlling cell death during metamorphosis in an hemimetabolous insect. *Dev. Biol.* **346**, 150–160.

Marie, B. and Bacon, J. P. (2000). Two *engrailed*-related genes in the cockroach: cloning, phylogenetic analysis, expression and isolation of splice variants. *Dev. Genes Evol.* **210**, 436–448.

Marie, B., Bacon, J. P. and Blagburn, J. M. (2000). Double-stranded RNA interference shows that Engrailed controls the synaptic specificity of identified sensory neurons. *Curr. Biol.* **10**, 289–292.

Marie, B., Cruz-Orengo, L. and Blagburn, J. M. (2002). Persistent *engrailed* expression is required to determine sensory axon trajectory, branching, and target choice. *J. Neurosci.* **22**, 832–841.

Marshall, J. L., Huestis, D. L., Hiromasa, Y., Wheeler, S., Oppert, C., Marshall, S. A., Tomich, J. M. and Oppert, B. (2009). Identification, RNAi knockdown, and functional analysis of an ejaculate protein that mediates a postmating, prezygotic phenotype in a cricket. *PLoS One* **4**, e7537.

Martín, D., Maestro, O., Cruz, J., Mané-Padrós, D. and Bellés, X. (2006). RNAi studies reveal a conserved role for RXR in molting in the cockroach *Blattella germanica*. *J. Insect Physiol.* **52**, 410–416.

McGavin, G. and Lewington, R. (2001). Essential Entomology: An Order-By-Order Introduction. Oxford University Press, Oxford.

Meyering-Vos, M. and Müller, A. (2007). RNA interference suggests sulfakinins as satiety effectors in the cricket *Gryllus bimaculatus*. *J. Insect Physiol.* **53**, 840–848.

Meyering-Vos, M., Merz, S., Sertkol, M. and Hoffmann, K. H. (2006). Functional analysis of the allatostatin-A type gene in the cricket *Gryllus bimaculatus* and the armyworm *Spodoptera frugiperda*. *Insect Biochem. Mol. Biol.* **36**, 492–504.

Mito, T., Sarashina, I., Zhang, H. J., Iwahashi, A., Okamoto, H., Miyawaki, K., Shinmyo, Y., Ohuchi, H. and Noji, S. (2005). Non-canonical functions of hunchback in segment patterning of the intermediate germ cricket *Gryllus bimaculatus*. *Development* **132**, 2069–2079.

Mito, T., Okamoto, H., Shinahara, W., Shinmyo, Y., Miyawaki, K., Ohuchi, H. and Noji, S. (2006). *Krüppel* acts as a gap gene regulating expression of *hunchback* and *even-skipped* in the intermediate germ cricket *Gryllus bimaculatus*. *Dev. Biol.* **294**, 471–481.

Mito, T., Kobayashi, C., Sarashina, I., Zhang, H., Shinahara, W., Miyawaki, K., Shinmyo, Y., Ohuchi, H. and Noji, S. (2007). *even-skipped* has gap-like, pair-rule-like, and segmental functions in the cricket *Gryllus bimaculatus*, a basal, intermediate germ insect (Orthoptera). *Dev. Biol.* **303**, 202–213.

Mito, T., Nakamura, T. and Noji, S. (2010). Evolution of insect development: to the hemimetabolous paradigm. *Curr. Opin. Genet. Dev.* **20**, 355–361.

Mito, T., Nakamura, T., Bando, T., Ohuchi, H. and Noji, S. (2011). The advent of RNA interference in entomology. *Entomological Science* **14**, 1–8.

Miyawaki, K., Mito, T., Sarashina, I., Zhang, H., Shinmyo, Y., Ohuchi, H. and Noji, S. (2004). Involvement of Wingless/Armadillo signaling in the posterior sequential segmentation in the cricket, *Gryllus bimaculatus* (Orthoptera), as revealed by RNAi analysis. *Mech. Dev.* **121**, 119–130.

Moriyama, Y., Sakamoto, T., Karpova, S. G., Matsumoto, A., Noji, S. and Tomioka, K. (2008). RNA interference of the clock gene *period* disrupts circadian rhythms in the cricket *Gryllus bimaculatus. J. Biol. Rhythms* **23**, 308–318.

Moriyama, Y., Sakamoto, T., Matsumoto, A., Noji, S. and Tomioka, K. (2009). Functional analysis of the circadian clock gene *period* by RNA interference in nymphal crickets *Gryllus bimaculatus. J. Insect Physiol.* **55**, 396–400.

Mutti, N. S., Park, Y., Reese, J. C. and Reeck, G. R. (2006). RNAi knockdown of a salivary transcript leading to lethality in the pea aphid, *Acyrthosiphon pisum. J. Insect Sci.* **6**, 1–7.

Mutti, N. S., Louis, J., Pappan, L. K., Pappan, K., Begum, K., Chen, M. S., Park, Y., Dittmer, N., Marshall, J., Reese, J. C. and Reeck, G. R. (2008). A protein from the salivary glands of the pea aphid, *Acyrthosiphon pisum*, is essential in feeding on a host plant. *Proc. Natl. Acad. Sci. USA* **105**, 9965–9969.

Nakamura, T., Mito, T., Tanaka, Y., Bando, T., Ohuchi, H. and Noji, S. (2007). Involvement of canonical Wnt/Wingless signaling in the determination of the positional values within the leg segment of the cricket *Gryllus bimaculatus. Dev. Growth Differ.* **49**, 79–88.

Nakamura, T., Mito, T., Miyawaki, K., Ohuchi, H. and Noji, S. (2008a). EGFR signaling is required for re-establishing the proximodistal axis during distal leg regeneration in the cricket *Gryllus bimaculatus* nymph. *Dev. Biol.* **319**, 46–55.

Nakamura, T., Mito, T., Bando, T., Ohuchi, H. and Noji, S. (2008b). Dissecting insect leg regeneration through RNA interference. *Cell. Mol. Life Sci.* **65**, 64–72.

Nakamura, T., Yoshizaki, M., Ogawa, S., Okamoto, H., Shinmyo, Y., Bando, T., Ohuchi, H., Noji, S. and Mito, T. (2010). Imaging of transgenic cricket embryos reveals cell movements consistent with a syncytial patterning mechanism. *Curr. Biol.* **20**, 1641–1647.

Ohde, T., Masumoto, M., Yaginuma, T. and Niimi, T. (2009). Embryonic RNAi analysis in the firebrat, *Thermobia domestica*: distal-less is required to form caudal filament. *J. Insect Biotechnol. Sericol.* **78**, 99–105.

Ott, S. R. and Rogers, S. M. (2010). Gregarious desert locusts have substantially larger brains with altered proportions compared with the solitarious phase. *Proc. Biol. Sci.* **277**, 3087–3096.

Panfilio, K. A. (2008). Extraembryonic development in insects and the acrobatics of blastokinesis. *Dev. Biol.* **313**, 471–491.

Panfilio, K. A. (2009). Late extraembryonic morphogenesis and its *zen*^{RNAi}-induced failure in the milkweed bug *Oncopeltus fasciatus. Dev. Biol.* **333**, 297–311.

Panfilio, K. A. and Roth, S. (2010). Epithelial reorganization events during late extraembryonic development in a hemimetabolous insect. *Dev. Biol.* **340**, 100–115.

Panfilio, K. A., Liu, P. Z., Akam, M. and Kaufman, T. C. (2006). *Oncopeltus fasciatus zen* is essential for serosal tissue function in katatrepsis. *Dev. Biol.* **292**, 226–243.

Pener, M. P. and Simpson, S. J. (2009). Locust phase polyphenism: an update. *Adv. Insect Physiol.* **36**, 1–286.

Piulachs, M. D., Pagone, V. and Bellés, X. (2010). Key roles of the *Broad-Complex* gene in insect embryogenesis. *Insect Biochem. Mol. Biol.* **40**, 468–475.

Pratt, G. E., Farnsworth, D. E., Siegel, N. R., Fok, K. F. and Feyereisen, R. (1989). Identification of an allatostatin from adult *Diploptera punctata. Biochem. Biophys. Res. Commun.* **163**, 1243–1247.

Price, D. and Gatehouse, J. (2008). RNAi-mediated crop protection against insects. *Trends Biotechnol.* **26**, 393–400.

Pueyo, J. I., Lanfear, R. and Couso, J. P. (2008). Ancestral Notch-mediated segmentation revealed in the cockroach *Periplaneta americana. Proc. Natl. Acad. Sci. USA* **105**, 16614–16619.

Regier, J. C., Shultz, J. W., Zwick, A., Hussey, A., Ball, B., Wetzer, R., Martin, J. W. and Cunningham, C. W. (2010). Arthropod relationships revealed by phylogenomic analysis of nuclear protein-coding sequences. *Nature* **463**, 1079–1083.

Revuelta, L., Piulachs, M. D., Bellés, X., Castañera, P., Ortego, F., Díaz-Ruíz, J. R., Hernández-Crespo, P. and Tenllado, F. (2009). RNAi against *ace1* and *ace2* in *Blattella germanica* reveals their differential contribution to acetylcholinesterase activity and sensitivity to insecticides. *Insect Biochem. Mol. Biol.* **39**, 913–919.

Roessingh, P., Simpson, S. J. and James, S. (1993). Analysis of phase-related changes in behaviour of desert locust nymphs. *Proc. Roy. Soc. London B: Biol. Sci.* **252**, 43–49.

Rogers, B. T., Peterson, M. D. and Kaufman, T. C. (2002). The development and evolution of insect mouthparts as revealed by the expression patterns of gnathocephalic genes. *Evol. Dev.* **4**, 96–110.

Romano, N. and Macino, G. (1992). Quelling: transient inactivation of gene expression in Neurospora crassa by transformation with homologous sequences. *Mol. Microbiol.* **6**, 3343–3353.

Ronco, M., Uda, T., Mito, T., Minelli, A., Noji, S. and Klingler, M. (2008). Antenna and all gnathal appendages are similarly transformed by *homothorax* knock-down in the cricket *Gryllus bimaculatus*. *Dev. Biol.* **313**, 80–92.

Rotenberg, D. and Whitfield, A. E. (2010). Analysis of expressed sequence tags for *Frankliniella occidentalis*, the western flower thrips. *Insect Mol. Biol.* **19**, 537–551.

Sakamoto, T., Uryu, O. and Tomioka, K. (2009). The clock gene *period* plays an essential role in photoperiodic control of nymphal development in the cricket *Modicogryllus siamensis*. *J. Biol. Rhythms* **24**, 379–390.

Saunders, D. S. (1990). The circadian basis of ovarian diapause regulation in *Drosophila melanogaster*: is the *period* gene causally involved in photoperiodic time measurement? *J. Biol. Rhythms* **5**, 315–331.

Shakesby, A. J., Wallace, I. S., Isaacs, H. V., Pritchard, J., Roberts, D. M. and Douglas, A. E. (2009). A water-specific aquaporin involved in aphid osmoregulation. *Insect Biochem. Mol. Biol.* **39**, 1–10.

Shinmyo, Y., Mito, T., Matsushita, T., Sarashina, I., Miyawaki, K., Ohuchi, H. and Noji, S. (2005). *caudal* is required for gnathal and thoracic patterning and for posterior elongation in the intermediate-germband cricket *Gryllus bimaculatus*. *Mech. Dev.* **122**, 231–239.

Shinmyo, Y., Mito, T., Uda, T., Nakamura, T., Miyawaki, K., Ohuchi, H. and Noji, S. (2006). *brachyenteron* is necessary for morphogenesis of the posterior gut but not for anteroposterior axial elongation from the posterior growth zone in the intermediate-germband cricket *Gryllus bimaculatus*. *Development* **133**, 4539–4547.

Sokol, N. S., Xu, P., Jan, Y. N. and Ambros, V. (2008). *Drosophila* let-7 microRNA is required for remodeling of the neuromusculature during metamorphosis. *Genes Dev.* **22**, 1591–1596.

Speight, M. R., Hunter, M. D. and Watt, A. D. (2008). Ecology of Insects: Concepts and Applications. Wiley-Blackwell, Oxford.

Stram, Y. and Kuzntzova, L. (2006). Inhibition of viruses by RNA interference. *Virus Genes* **32**, 299–306.

Suazo, A., Gore, C. and Schal, C. (2009). RNA interference-mediated knock-down of *Bla g 1* in the German cockroach, *Blattella germanica* L., implicates this allergen-encoding gene in digestion and nutrient absorption. *Insect Mol. Biol.* **18**, 727–736.

Sutherland, T. D., Young, J. H., Weisman, S., Hayashi, C. Y. and Merritt, D. J. (2010). Insect silk: one name, many materials. *Annu. Rev. Entomol.* **55**, 171–188.

Takahashi, T., Hamada, A., Miyawaki, K., Matsumoto, Y., Mito, T., Noji, S. and Mizunami, M. (2009). Systemic RNA interference for the study of learning and memory in an insect. *J. Neurosci. Methods* **179**, 9–15.

Thakker, D. R., Hoyer, D. and Cryan, J. F. (2006). Interfering with the brain: use of RNA interference for understanding the pathophysiology of psychiatric and neurological disorders. *Pharmacol. Ther.* **109**, 413–438.

Timmons, L. and Fire, A. (1998). Specific interference by ingested dsRNA. *Nature* **395**, 854.

Timmons, L., Court, D. L. and Fire, A. (2001). Ingestion of bacterially expressed dsRNAs can produce specific and potent genetic interference in *Caenorhabditis elegans*. *Gene* **263**, 103–112.

Tobback, J., Boerjan, B., Vandersmissen, H. P. and Huybrechts, R. (2011). The circadian clock genes affect reproductive capacity in the desert locust Schistocerca gregaria. *Insect Biochem. Mol. Biol.* **41**, 313–321.

Tomioka, K., Sakamoto, T. and Moriyama, Y. (2009). RNA interference is a powerful tool for chronobiological study in the cricket. *Sleep and Biological Rhythms* **7**, 144–151.

Truman, J. W. and Riddiford, L. M. (1999). The origins of insect metamorphosis. *Nature* **410**, 447–452.

Truman, J. W. and Riddiford, L. M. (2007). The morphostatic actions of juvenile hormone. *Insect Biochem. Mol. Biol.* **37**, 761–770.

Turner, C. T., Davy, M. W., MacDiarmid, R. M., Plummer, K. M., Birch, N. P. and Newcomb, R. D. (2006). RNA interference in the light brown apple moth, *Epiphyas postvittana* (Walker) induced by double-stranded RNA feeding. *Insect Mol. Biol.* **15**, 383–391.

Voinnet, O. (2001). RNA silencing as a plant immune system against viruses. *Trends Genet.* **17**, 449–459.

Wang, Q. L. and Li, Z. H. (2007). The functions of microRNAs in plants. *Front. Biosci.* **12**, 3975–3982.

Wei, Z., Baggerman, G., Nachman, R. J., Goldsworthy, G., Verhaert, P., De Loof, A. and Schoofs, L. (2000). Sulfakinins reduce food intake in the desert locust, *Schistocerca gregaria*. *J. Insect Physiol.* **46**, 1259–1265.

Wei, Z., Yin, Y., Zhang, B., Wang, Z., Peng, G., Cao, Y. and Xia, Y. (2007). Cloning of a novel protease required for the molting of *Locusta migratoria manilensis*. *Dev. Growth Differ.* **49**, 611–621.

Whyard, S., Singh, A. D. and Wong, S. (2009). Ingested double-stranded RNAs can act as species-specific insecticides. *Insect Biochem. Mol. Biol.* **39**, 824–832.

Zhang, J., Liu, X., Zhang, J., Li, D., Sun, Y., Guo, Y., Ma, E. and Zhu, K. Y. (2010). Silencing of two alternative splicing-derived mRNA variants of chitin synthase 1 gene by RNAi is lethal to the oriental migratory locust, *Locusta migratoria manilensis* (Meyen). *Insect Biochem. Mol. Biol.* **40**, 824–833.

Zhao, Y. and Srivastava, D. (2007). A developmental view of microRNA function. *Trends Biochem. Sci.* **32**, 189–197.

Zhou, X. and Riddiford, L. (2002). Broad specifies pupal development and mediates the 'status quo' action of juvenile hormone on the pupal-adult transformation in *Drosophila* and *Manduca*. *Development* **129**, 2259–2269.

Zhou, X. and Xia, Y. (2009). Cloning of an acetylcholinesterase gene in *Locusta migratoria manilensis* related to organophosphate insecticide resistance. *Pestic. Biochem. Physiol.* **93**, 77–84.

Zhou, X., Oi, F. M. and Scharf, M. E. (2006a). Social exploitation of hexamerin: RNAi reveals a major caste-regulatory factor in termites. *Proc. Natl. Acad. Sci. USA* **103**, 4499–4504.

Zhou, X., Tarver, M. R., Bennett, G. W., Oi, F. M. and Scharf, M. E. (2006b). Two hexamerin genes from the termite *Reticulitermes flavipes*: sequence, expression, and proposed functions in caste regulation. *Gene* **376**, 47–58.

Zhou, X., Tarver, M. R. and Scharf, M. E. (2007). Hexamerin-based regulation of juvenile hormone-dependent gene expression underlies phenotypic plasticity in a social insect. *Development* **134**, 601–610.

Zhou, X., Wheeler, M. M., Oi, F. M. and Scharf, M. E. (2008). RNA interference in the termite *Reticulitermes flavipes* through ingestion of double-stranded RNA. *Insect Biochem. Mol. Biol.* **38**, 805–815.

New Approaches to Agricultural Insect Pest Control Based on RNA Interference

Xue-Yi Xue,* Ying-Bo Mao,* Xiao-Yuan Tao,* Yong-Ping Huang,† and Xiao-Ya Chen*

*National Key Laboratory of Plant Molecular Genetics, Institute of Plant Physiology and Ecology, Shanghai Institutes for Biological Sciences, Chinese Academy of Sciences, Shanghai, PR China
†CAS Key Laboratory of Entomology, Institute of Plant Physiology and Ecology, Shanghai Institutes for Biological Sciences, Chinese Academy of Sciences, Shanghai, PR China

Abstract

In the past decade, RNA interference (RNAi) has been in rapid development as a powerful tool in functional genomics research. The discovery of RNAi has augmented our understanding of small (~20–30 nt) noncoding RNAs as critical regulators of gene

ADVANCES IN INSECT PHYSIOLOGY VOL. 42
ISBN 978-0-12-387680-5
DOI: 10.1016/B978-0-12-387680-5.00003-3

expression and genome stability, and the RNAi also presents a fascinating technology with great potential for application in gene function dissection, disease therapy, plant protection, and nutritional improvement, as well as many other areas. In this review, we give a brief introduction to the discovery of RNAi, describe its molecular mechanisms and discuss the biological functions of small RNAs in insects and the development of new technologies for insect pest control. Emphasis is placed on methods of double-stranded RNA (dsRNA) delivery and pathways of dsRNA uptake by insects, especially the engineering of plants to specifically suppress herbivorous insect gene expression. We conclude by considering feasible approaches for enhancing the effectiveness of RNAi and potential drawbacks of this promising approach to agricultural pest control.

1 Introduction

1.1 CURRENT PROBLEMS IN PEST CONTROL

Insect pests cause severe economic losses in crop production every year. In the past half-century, increasing grain yield has relied heavily on more efficacious pest control through the use of synthetic chemical pesticides (SCPs). About 60 years ago, when organic pesticides became popular in agriculture and forestry, they were thought to represent the final battle with pests. However, the situation was not as simple as initially expected. Many problems associated with the application of pesticides have shown up, including environment pollution, harmful side effects on non-target organisms, resurgence of pest populations due to the disruption of natural control, the evolution of resistance by target pests and increasing cost. These problems will persist unless equal or even more effective alternative technologies are developed.

In the past half-century, particularly since the publication of the 'Silent Spring' by Rachel Carson in 1962, attention has been paid to developing new pest control strategies, and to replacing or reducing the wide application of SCPs. These efforts include the use of natural enemies, natural product-based pesticides, pest-resistant genetically modified (GM) crops and combinations of these strategies. Among the limited successful examples, the insecticidal crystal proteins produced by *Bacillus thuringiensis* (Bt) have been the most promising alternative to synthetic insecticides. Bt insecticides kill certain groups of major insect pests, such as mosquitoes and crop-eating caterpillars (lepidopteran larvae), but are harmless to vertebrates and most other organisms (de Maagd *et al.*, 2001). Bt series have been applied to many crops through conventional sprayable formulations or by transgenic approaches to control lepidopteran insect pests. To date, the most widely used Bt insecticides are from the Cry1A crystal protein family, particularly Cry1Ab in transgenic Bt corn and Cry1Ac in transgenic Bt cotton, which kill caterpillars (Tabashnik *et al.*, 2003). In China, Bt cotton was approved for commercial use in 1997; since then it has shown a great success in controlling the cotton bollworm, *Helicoverpa armigera*, and in

2010 the biotech cotton accounted for 95% of cotton grown in Northern China (Lu *et al.*, 2010).

However, the development of insect resistance to Bt toxins observed in laboratory and field populations exposed to transgenic plants poses a threaten to the current strategy of pest control (Bagla, 2010; Tabashnik *et al.*, 2003, 2008). At least seven resistant strains of three pests (*Plutella xylostella*, *Pectinophora gossypiella*, and *H. armigera*) have been isolated in laboratory and greenhouse tests (Tabashnik *et al.*, 2003). Furthermore, Bt resistance in field populations of *Helicoverpa zea* in America (Tabashnik *et al.*, 2008) and pink bollworm (*P. gossypiella*) fed on *Cry1Ac* cotton in India (Bagla, 2010) have also been reported. In addition to the evolution of resistance, the composition of pest populations has been found to shift from a single dominant species to a more diverse community of pest species as a result of Bt exposure. A recent report illustrated that wide-scale cultivation of Bt cotton in China to some extent resulted in large outbreaks of another type of herbivorous insects, the mirid bugs (Heteroptera: Miridae), on multiple crops (Lu *et al.*, 2010). A meta-analysis similarly showed that non-target invertebrates were generally more abundant in Bt cotton and Bt maize fields than in non-transgenic fields managed with insecticides (Marvier *et al.*, 2007), though this may be a good news from ecological points of view.

1.2 NEW TRENDS IN PEST CONTROL

Facing such problems, new and environmentally friendly pest control technologies have become a necessity. Ideal pest control methods would have the following characteristics: species specificity, absence of side effects on crops and no or negligible environmental pollution. RNAi and related techniques show such potential. Investigations of insects, particularly the model fruit fly, *Drosophila melanogaster*, have contributed greatly to elucidation of RNAi mechanisms (Brennecke *et al.*, 2007; Czech *et al.*, 2008; Gunawardane *et al.*, 2007; Liu *et al.*, 2003; Okamura *et al.*, 2008b; Saito *et al.*, 2006; Yin and Lin, 2007) and RNAi is now widely used as a powerful tool for research in insect functional genomics. The insect genomics research has been providing an increasingly huge amount of data. As of January 2012, 109 insect genomes were either finished or in progress according to NCBI Entrez Genome Project Database, and most of them are publicly available, which means that there will be more genes with unknown functions that await investigation. Unlike plants, especially those with self-pollination, for which mutant libraries can be generated and preserved with a relatively high efficiency, usually tedious work is required for insect reverse genetics. Therefore, an efficient RNAi platform is invaluable for entomological research. Furthermore, because RNAi selectively inhibits target gene expression with a high specificity and fidelity, it is naturally a strategy of choice to be used in controlling insect pests, as long as a delivery method is available. In 2007, two papers, including one of our own, reported the use of plant-mediated RNAi to suppress the growth of two agricultural pests, the cotton bollworm *H. armigera* (Mao *et al.*, 2007) and the western corn rootworm (WCR) *Diabrotica virgifera virgifera* LeConte (Baum *et al.*, 2007). The breakthrough highlighted the potential for further development of the RNAi technology for crop protection. For RNAi to be broadly successful as a pest control approach, the following have to be taken into account: synthesis of double-stranded RNA (dsRNA), entry of dsRNA into pest cells, mechanisms of spread of the silencing signal, efficiency and persistence of the silencing effect and biosafety evaluation, as well as potential shortcomings and countermeasures. Below, we address each of these items in turn.

2 RNA interference and the RNAi machinery

To better understand this new pest control strategy, we begin by briefly introducing the history of RNAi research and key proteins involved in the RNAi response, Dicers and Argonautes (AGO), in different species. Incidentally, general classification of small RNAs and functions of insect small RNAs are discussed.

2.1 DISCOVERY OF RNA INTERFERENCE

In 1990, to determine whether chalcone synthase (CHS) was a key enzyme in flavonoid biosynthesis, Jorgensen and colleagues overexpressed a CHS gene (*chsA*) under the control of a CaMV 35S promoter in petunia (*Petunia hybrida* L.). Unexpectedly, up to half of the transgenic lines showed white petunias, rather than the expected deepening violet colour (Napoli *et al.*, 1990), and the transcript levels of endogenous as well as introduced *CHS* genes were dramatically reduced. Because the introduced transgene triggered the reduction of endogenous *CHS* gene expression, this phenomenon was called co-suppression (Angenent *et al.*, 1994; Brusslan *et al.*, 1993; De Carvalho *et al.*, 1992; Matzke and Matzke, 1991). In fact, before co-suppression was proposed, a similar phenomenon had been documented in the 1980s. To test the hypothesis of 'use a portion of a pathogen's own genetic material against itself' proposed by Sanford and Johnston (1985), Beachy and colleagues introduced the coat protein (CP) gene of the tobamovirus *tobacco mosaic virus* (TMV) into tobacco. They found that, after incubation with TMV, transgenic plants either did not display symptoms or showed delayed symptoms of infection. They presented this phenomenon as coat protein-mediated resistance (CPMR) (Abel *et al.*, 1986).

RNA silencing was first noticed by Guo and Kemphues (1995) in the nematode *Caenorhabditis elegans*. At that time, antisense RNAs were being tested for their ability to down-regulate target genes. However, treatment of nematodes with sense RNA of *par-1* also resulted in suppressed *par-1* expression.

This paradoxical finding was explained by Fire *et al.* (1998), who demonstrated that gene silencing was triggered by dsRNA, not by single-stranded RNA. This discovery allowed links to be made between the phenomena of co-suppression, CPMR and sense RNA suppression. Subsequent investigations into the detection of the antisense strand in co-suppression plants (Hamilton and Baulcombe, 1999), the mechanism of post-transcriptional gene silencing (PTGS) (Hammond *et al.*, 2000; Zamore *et al.*, 2000), the discovery of RNase III enzymes (Bass, 2000; Catalanotto *et al.*, 2000; Cerutti *et al.*, 2000; Martinez *et al.*, 2002; Tabara *et al.*, 1999) and the length of miRNAs (Reinhart *et al.*, 2000) opened a new field of genetic and epigenetic regulation.

2.2 RNAI MACHINERY

Although many classes of small RNAs have been identified and there are some distinctions between different RNAi pathways, the core components of each pathway are shared and conserved. Small RNAs have been classified based on the structure of their precursors, with three major classes: short interfering RNAs (siRNAs), microRNAs (miRNAs) and Piwi-interacting RNAs (piRNAs) (Table 1). The siRNAs, ~20–25 nt, act as defenders of genome integrity in response to foreign or invasive nucleic acids, and are derived from long dsRNA molecules that come from replication of RNA viruses, convergent repeat mobile elements, endogenous invert repeats, self-annealing transcripts or experimental dsRNA transgenes. By contrast, miRNAs are transcribed from the organism's own genome and regulate endogenous gene expression. The piRNAs, which are ~24–31 nt long, differ from canonical siRNAs and miRNAs in several aspects, and they silence transposons and retrotransposons in animal germ cells (Lin, 2007).

RNAi refers to the RNA-based sequence-specific suppression of cognate gene expression. It can occur at both the transcriptional and the post-transcriptional levels. The precursor production pathways of different small RNAs can differ, but the processing and effect steps have a common mode in fungi, plants, worms, insects and vertebrates. Although diverse proteins are involved in production of small RNAs, Dicers and Argonautes are core components of the small RNA molecule production and the RNAi effect steps, respectively. Table 2 shows the Dicers and Argonautes in different model species and subclades of Argonaute proteins in animals. Dicer RNase III family proteins are involved in the production of small RNA molecules and have several conserved motifs, including an amino-terminal DExH-Box helicase domain, a Piwi/Argonaute/Zwille (PAZ) domain, two RNase III motifs and a carboxy-terminal dsRNA binding domain (*Drosophila* Drasha lacks the RNA helicase and PAZ domains). Because this review article is focused on RNAi effects related to plant–insect interactions, the small RNA biosynthesis pathways and the RNAi mechanisms will be discussed for both insects and plants.

TABLE 1
Small RNA species

Small RNA		Length (nt)	Origin	Silence mechanisms	Organisms	References
siRNA class	siRNA[a]	~21–25	Transgene, virus, endogenous repeat-associated elements	PTGS, CDGS	Fungi, plants, animals	Grosshans and Filipowicz (2008), Moazed (2009), Dunoyer *et al*. (2010)
	nat-siRNA	~20–25	Natural *cis*-antisense transcripts (NAT)	PTGS	Plants, animals	Katiyar-Agarwal *et al*. (2006), Okamura *et al*. (2008a)
	ta-siRNA	21	TAS genes	PTGS	Plants	Vazquez *et al*. (2004), Allen *et al*. (2005)
	lsiRNAs	~30–40	NAT pair	PTGS	Plants	Katiyar-Agarwal *et al*. (2007)
miRNA		~21–25	MiRNA encoding genes	PTGS, CDGS	Plants, animals	Carthew and Sontheimer (2009), Moazed (2009)
piRNA		~24–31 or 21[b]	Repetitive DNA, transposons	PTGS, CDGS	Animals (germ cells)	Hartig *et al*. (2007), Lin (2007), Moazed (2009)

[a] Plant and worm endo-siRNAs are typically produced through RNA-dependent RNA polymerase (RdRP). More recently, endo-siRNAs have been detected in somatic and germ cells of *Drosophila* species and in mouse oocytes. These animal endo-siRNAs are sorted as general siRNAs because they originate from endogenous dsRNAs precursors (Ghildiyal and Zamore, 2009).

[b] *C. elegans* piRNAs called 21U-RNAs are 21 nt long (Batista *et al*., 2008; Das *et al*., 2008).

RNAi pathways identified up to now seem to act in both post-transcriptional gene silencing (PTGS) and chromatin-dependent gene silencing (CDGS) pathways. *Trans*-acting siRNAs (ta-siRNAs) and long siRNAs (lsiRNAs) are plant-specific endo-siRNAs.

TABLE 2
Dicer and Argonaute in different organisms

Proteins		Yeast (*Schizosaccharomyces pombe*)	Plant (*Arabidopsis thaliana*)	Nematode (*Caenorhabditis elegans*)[a]	Fruitfly (*Drosophila melanogaster*)	Zebrafish (*Danio rerio*)	Human (*Homo sapiens*)
Dicer family		Dcr1	DCL1 DCL2 DCL3 DCL4	DCR-1 DRSH-1	Dcr-1 Dcr-2 Drosha	Dicer1 Drosha	Dicer1 Drosha
Argonaute family	AGO clade	Ago1	AGO1 AGO2 AGO4 AGO6 AGO7 AGO9 AGO10 [b]	ALG-1 ALG-2 ALG-3 ALG-4 CeT23D8.7	Ago1 Ago2	Ago1 Ago2 Ago3 Ago4	AGO1 AGO2 AGO3 AGO4
	PIWI clade	None	None	ERGO-1 PRG-1 PRG-2	Ago3 Piwi Aub	ZIWI ZILI	HILI HIWI HIWI2 HIWI3

[a] There is a striking expansion of Argonaute proteins in *C. elegans* and 27 Argonaute proteins have been identified (Yigit *et al*., 2006). The AGO subfamily contains five members. Function of the fifth AGO protein (CeT23D8.7) is unknown (Conine *et al*., 2010). In addition to the AGO and PIWI subfamilies in *C. elegans*, an expanded worm-specific Argonaute subfamily, WAGO, plays important roles in the silencing response triggered by secondary siRNAs (Siomi and Siomi, 2009). This Argonaute subfamily is not listed here.

[b] Ten conserved AGOs have been identified in *Arabidopsis*. Three AGOs belonging to this family but without a known function in RNAi are not listed here.

Data were obtained from the references Yigit *et al*. (2006), Chapman and Carrington (2007), Houwing *et al*. (2007), Hutvagner and Simard (2008), Tomoyasu *et al*. (2008), Cifuentes *et al*. (2010).

Regarding Dicer proteins, the number of dicers varies across species. There is only one Dicer protein in mouse, human and *C. elegans* (Table 2), whereas in *Drosophila*, two Dicer proteins are found, which associate with the siRNA and miRNA pathways, respectively (Lee *et al.*, 2004). In the model plant *Arabidopsis thaliana*, four Dicer-like genes, *DCL1*, *DCL2*, *DCL3* and *DCL4*, are present. DCL1 catalyses miRNA biogenesis, while DCL2, DCL3 and DCL4 are involved in the production of different types of siRNAs. DCL2 is mainly involved in processing of natural antisense transcript-derived siRNAs (nat-siRNAs) and *trans*-acting siRNA (ta-siRN) precursors (Gasciolli *et al.*, 2005). DCL3 is the main player in the production of 24-nt repeat-associated siRNA (ra-siRNA) essential for transcriptional gene silencing and for establishing and maintaining DNA methylation (Chan *et al.*, 2004; Onodera *et al.*, 2005). DCL4 is an antiviral Dicer producing 21nt virus-derived siRNAs. Furthermore, the 22-nt ra-siRNAs generated by DCL2 and 21-nt ra-siRNAs generated by DCL4 are sufficient to maintain certain DNA methylation marks in *dcl3* mutants (Henderson *et al.*, 2006). Hence, in plants DCL proteins have major functions in specific pathways but functional redundancy exists between members (Vazquez, 2006).

Argonaute proteins are key components of the RNA-induced silencing complex (RISC) and some are also involved in transcriptional silencing in the RNA-induced transcriptional silencing complex (RITS) (Lippman and Martienssen, 2004; Verdel *et al.*, 2004). Argonaute proteins contain three characteristic domains: amino-terminal PAZ, carboxy-terminal Piwi and Mid (Song and Joshua-Tor, 2006). The PAZ domain appears to be an RNA-binding domain (Lingel *et al.*, 2003; Song *et al.*, 2003); the Piwi domain, which has structural and functional similarity to RNase H, cleaves target RNAs through its endonucleolytic activity (Song *et al.*, 2004). Different Argonaute proteins are recruited to different processes. *Drosophila* has five Argonaute proteins (Table 2). Ago1 and Ago2 are predominantly involved in the miRNA and siRNA pathways, respectively (Okamura *et al.*, 2004), while Piwi, Aubergine (Aub) and Ago3 are used in transcriptional gene silencing by piRNAs (Brennecke *et al.*, 2007; O'Donnell and Boeke, 2007; Pal-Bhadra *et al.*, 2004). The PIWI clade of the Argonaute protein family is found in all animals examined so far, and its presence is tightly correlated with the emergence of specialized germ cells (Aravin *et al.*, 2007). Like *Drosophila*, *Tribolium castaneum* (Coleoptera) also has five Argonaute genes and these AGO proteins are predicted to perform the same functions as their *Drosophila* homologues (Tomoyasu *et al.*, 2008). The *Arabidopsis* genome encodes 10 AGOs. AGO1 and AGO10 play a predominant role in miRNA function (Mallory *et al.*, 2009); AGO4, AGO6 and AGO9 function in ra-siRNA generation, RNA-directed DNA methylation, and transcriptional gene silencing of transposons and repeat-associated elements (Havecker *et al.*, 2010; Qi *et al.*, 2006; Zheng *et al.*, 2007; Zilberman *et al.*, 2003); AGO7 appears to be associated with ta-siRNA function in regulating developmental timing (Fahlgren *et al.*, 2006; Vazquez *et al.*, 2004). A recent study proposed that AGO2 plays an antiviral role, possibly by binding viral siRNAs that target the viral RNAs (Harvey *et al.*, 2011). Whether other *Arabidopsis* AGOs (AGO3, AGO5 and AGO8) participate in small RNA-directed regulation awaits further investigation.

To address why *Arabidopsis* possess functionally redundant AGOs, Mi and colleagues (2008) catalogued small RNAs associated with four *Arabidopsis* AGOs by deep sequencing. They documented that each AGO recruits a particular subset of small RNAs by recognizing their 5′-terminal nucleotides. For example, AGO2 and AGO4 preferentially recruit small RNAs with a 5′-terminal adenosine, whereas AGO1 harbours miRNAs with a 5′-terminal uridine. AGO5 predominantly selects small RNAs with 5′-cytosine (Mi *et al.*, 2008). These findings demonstrate that small RNAs with different 5′-terminal nucleotides are differentially recruited by different AGOs in *Arabidopsis*. However, in *D. melanogaster*, the identity of the 5′-nucleotides only makes a minor contribution, and thermodynamic properties also influence small RNA sorting (Czech and Hannon, 2011; Förstemann *et al.*, 2007). In general, Ago1 selects miRNAs, whereas Ago2 associates with siRNAs; this is coincident with the processing of miRNAs by Dcr-1 and siRNAs by Dcr-2 (Czech *et al.*, 2009; Okamura *et al.*, 2009).

2.3 BIOLOGICAL FUNCTIONS OF SMALL RNAS IN INSECTS

Before discussing the use of RNAi as a tool, we first illustrate the roles of endogenous small RNAs in insects. Since the discovery of the first miRNAs, *lin-4* and *let-7*, in *C. elegans* (Reinhart *et al.*, 2000), many miRNAs in insect species have been found by high throughput sequencing and bioinformatic analyses of insect genomes (Cai *et al.*, 2010; Chen *et al.*, 2010; Jagadeeswaran *et al.*, 2010; Legeai *et al.*, 2010; Li *et al.*, 2009; Sandmann and Cohen, 2007; Yu *et al.*, 2009). Insect small RNAs have been shown to participate in multiple biological processes, including embryogenesis, tissue differentiation, metamorphosis, circadian rhythm, insect defence and transposon activity suppression.

2.3.1 *miRNAs in insect development*

Forward and reverse genetic studies have successfully uncovered functions of a number of insect miRNAs, though these investigations have been limited mostly to *D. melanogaster*. The majority of miRNAs have more than one target gene, so that they have multiple roles in developmental pathways. For example, *Drosophila miR-184* controls germline stem cell differentiation, axis formation of the egg chamber and cellularization of embryo by regulating three different factors, Saxophone, gurken transport factor K10 and Tramtrack69, respectively (Iovino *et al.*, 2009). *Drosophila miR-9a* ensures the precise production of sensory organ precursors (SOPs) in the peripheral nervous system by targeting the transcription factor *senseless* (Li *et al.*, 2006) and represses *LIM-only* to prevent apoptosis during wing development (Bejarano *et al.*, 2009; Biryukova *et al.*, 2009).

2.3.2 *RNAi in antiviral defence*

In addition to regulating host cellular functions, the miRNA pathway has more recently been shown to play an important role in host–virus interactions (Ulvila *et al.*, 2010; see Chapter 1). Ascoviruses (AV) cause lethal infection in lepidopteran insects. By analysing the expression profile of cellular miRNAs using an insect cell line derived from *H. zea* fat bodies (HzFB) upon infection with *Heliothis virescens* AV (HvAV-3e), it was documented that *Hz-miR24* down-regulates ascoviral DNA-dependent RNA polymerase II RPC2 (*ORF64*) and its β subunit (*ORF82*) transcript levels late in infection; *Hz-miR24* is the first miRNA reported to interact with an insect virus (Hussain and Asgari, 2009).

2.3.3 *piRNAs and suppression of transposon activity*

Studies of transposon suppression in *Drosophila* Piwi mutants have pointed to a conserved function of piRNAs in the control of mobile genetic elements (Aravin *et al.*, 2001; Vagin *et al.*, 2006). In *Drosophila*, the *flamenco* locus is a piRNA cluster, which represses transposition of the retrotransposons *gypsy*, *ZAM* and *Idefix* (Desset *et al.*, 2003; Prud'homme *et al.*, 1995). The *flamenco* mutation resulted in defects in germ and follicle cell development and complete sterility. It was found that in *flamenco* mutants, all piRNAs derived from this locus were lost and *gypsy* was activated (Brennecke *et al.*, 2007).

3 Methodological considerations in using RNAi in insect

3.1 DSRNA PRODUCTION

3.1.1 *dsRNA synthesized* in vitro

The preferred delivery methodology in the majority of entomology studies is microinjection of long dsRNA, synthesized *in vitro*, into the insect haemocoel. Using a commercial *in vitro* transcription kit, dsRNAs are normally synthesized in laboratories. The general principle is as follows. The templates for *in vitro* transcription are prepared by PCR using gene-specific primers with the T7 polymerase promoter sequence at their 5′-end. The reaction mixture is incubated for several hours at 37°C with T7 polymerase, treated with DNase, and purified by standard methods. To maximize the yield of dsRNA, it is advisable to add an annealing step.

3.1.2 *dsRNAs produced by bacteria*

dsRNA may also be transcribed in bacteria rather than *in vitro*. *C. elegans* normally eat bacteria, ingesting and grinding them in the pharynx and subsequently absorbing bacterial contents in the gut. To silence nematode gene expression, Timmons and Fire engineered *Escherichia coli* to express dsRNAs directed against *C. elegans* genes, and found that bacteria expressing corresponding dsRNAs could confer specific interference effects to the nematode larvae fed on them (Timmons and Fire, 1998). One potential advantage of this approach is that dsRNA continuously produced in organisms is more stable than dsRNA transcribed *in vitro* when placed in food (Whyard *et al.*, 2009). Since then, this bacteria-mediated RNAi approach has been successfully applied to other organisms, including a planarian *Schmidtea mediterranea* (Newmark *et al.*, 2003) and the lepidopteran insect *Spodoptera exigua* (Tian *et al.*, 2009). It is encouraging that bacteria-mediated insect RNAi provides investigators an alternative platform for large-scale screening of candidate target genes.

3.1.3 *dsRNAs expressed in plants*

The dsRNAs mentioned above are synthesized *in vitro* or produced in bacteria. These dsRNA molecules are incorporated into an artificial diet or directly injected into insects with a microinjector, limiting the utilization of this approach to the laboratory. To exploit the feasible and practical benefit of RNAi in field control of pests, plants are a good choice for dsRNA production. First, plants are the host and food source of herbivorous insects. Second, plants have tons of biomass and could accumulate a large amount of dsRNAs to provoke the RNAi response. And third, dsRNAs can be continuously produced under varying environmental conditions. The dsRNAs could be produced in plants under universal or tissue-specific promoters, as well as under constitutive or inducible promoters. The possibility of using RNAi to protect plants against insect pests by down-regulating essential gene functions in the herbivore, thus resulting in its death or growth retardation, has been recognized for many years. However, this approach was considered unfeasible, because the absence of RNA-dependent RNA polymerase (RdRP) in insects implied that silencing effects produced by feeding dsRNAs to insects would be limited (see review by Price and Gatehouse, 2008). This view was challenged by the emergence of engineered plants expressing dsRNAs which conferred protection against nematodes (Huang *et al.*, 2006) and completely broken by the achievement of insect resistance via plant producing dsRNAs (Baum *et al.*, 2007; Mao *et al.*, 2007).

3.2 DSRNA DELIVERY

3.2.1 *Incubation or soaking*

To determine whether RNAi functions in cell culture, Dixon and colleagues compared expression profiles of three components of a signalling complex, DSH3PX1, DACK and Dock, in untreated *D. melanogaster* Schneider 2 (S2) cells and the S2 cells soaked in dsRNA solution. Western blots of cell extracts showed that *DSH3PX1* dsRNA specifically reduced the DSH3PX1 protein levels in a concentration-dependent manner, whereas *DACK* dsRNA similarly reduced the DACK protein level, which demonstrated that dsRNAs corresponding to *DSH3PX1* and *DACK*, respectively, caused the selective suppression of the respective proteins in S2 cell cultures. Since then, soaking cells in dsRNA-containing solutions has been effective against a broad range of genes in several *Drosophila* cell lines (Clemens *et al.*, 2000). In addition, it is also feasible to induce RNAi by soaking nematodes (Tabara *et al.*, 1998) and flatworms (Orii *et al.*, 2003) in dsRNA-containing solutions.

3.2.2 *Stable transformation*

In genetically transformable species, RNAi can be triggered by expression of a long double-stranded hairpin RNA from a transgene cascade containing an inverted repeat fragment. After achievement of heritable and inducible genetic interference by dsRNA in *C. elegans*, a strategy for transgenic RNAi in *D. melanogaster* was described, and the binary GAL4/UAS transgenic RNAi system is shown in Figure 1 (Kennerdell and Carthew, 2000). It was found that *lacZ* expression was inhibited by the presence of UAS-*IRlacZ* in F1 progeny. Gene suppression can be induced at any time of the life cycle using the binary GAL4/UAS system, which allows investigators to accurately disturb gene function at any stage and in a tissue-specific manner. Dickson and colleagues used GAL4/UAS system to generate and validate a genome-wide library of *D. melanogaster* RNAi transgenes, enabling the conditional inactivation of gene function in specific tissues of the intact organism. They generated 22,270 transgenic lines, covering 88% of the predicted protein-coding genes in the *Drosophila* genome (Dietzl *et al.*, 2007). This stable transgenic RNAi library makes it possible for systematic analysis of gene functions in any tissue and at any developmental stage of *Drosophila*. High levels of gene silencing were also documented in two studies of transgenic *B. mori* expressing dsRNAs (Dai *et al.*, 2008; Kanginakudru *et al.*, 2007).

3.2.3 *Injection*

Among the several methods that have been used for the delivery of dsRNA molecules into insects, direct microinjection has been the most popular one because of its simple protocol and effectiveness in silencing the expression of target genes. Shortly after the finding that dsRNAs are RNAi triggers (Fire *et al.*, 1998), RNAi was used to dissect gene function *in vivo* in *D. melanogaster* by dsRNA injection (Kennerdell and Carthew, 1998). To determine whether dsRNA-mediated interference could occur in *Drosophila*, the authors chose the previously characterized genes of *fushi tarazu* (*ftz*) and *even-skipped* (*eve*). They injected dsRNA into syncytial blastoderm embryos. Injection of either *ftz*- or *eve*-annealed RNA into wild-type embryos effectively interfered with the gene activity, as demonstrated by cuticle phenotypes characteristic of *ftz* or *eve*

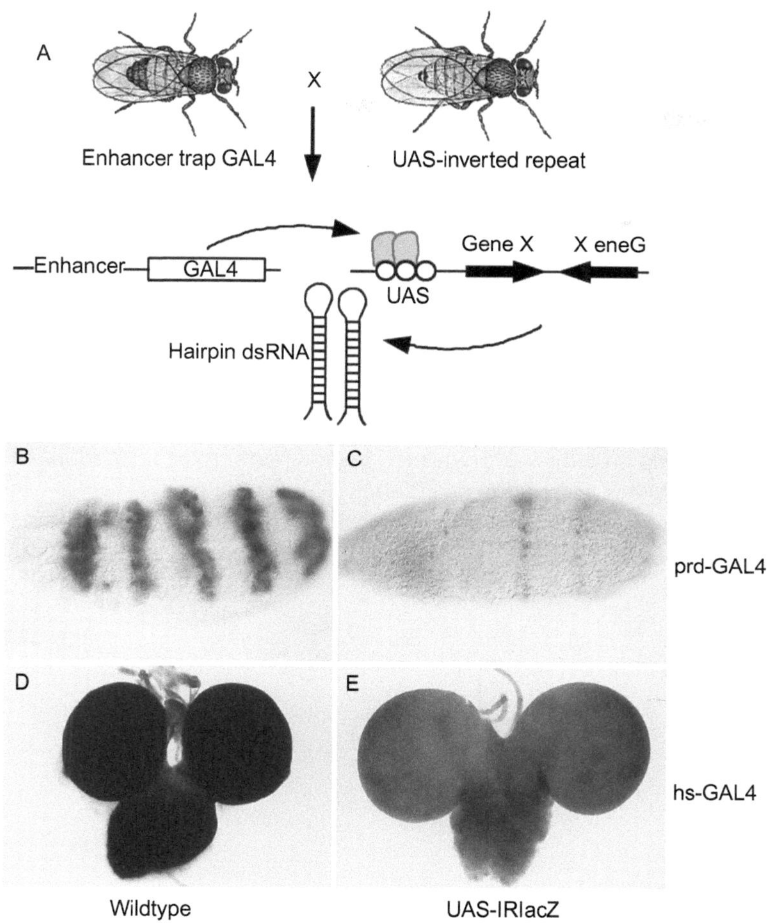

FIG. 1 *Drosophila melanogaster* RNAi transgene system. (A) Strategy for generation of transgenic RNAi. A fragment of a gene is dimerized to produce a long double-stranded hairpin RNA. The transgene is under the control of a generic promoter containing the GAL4-responsive upstream activator sequence (UAS) element in an inducible expression vector (right fly). Upon mating UAS-*dsRNA* transgenic flies with flies carrying a tissue- or cell-specific GAL4 transgene (left fly), the F1 progeny generated a hairpin-like dsRNA in all cells that expressed the GAL4 gene and in these cells the target genes were silenced (Kennerdell and Carthew, 2000). (B and D) F1 progeny carrying a GAL4 transgene and reporter UAS-*lacZ* gene but no UAS-*IRlacZ* transgene. (C and E) F1 progeny carrying the UAS-*IRlacZ* that was predicted to make lacZ dsRNA. (B and C) Stage 10 embryos carrying paired-GAL4 showing a pair-rule pattern of lacZ expression that was inhibited by the presence of UAS-IRlacZ. (D and E) Brain complex and thoracic nerve cord from late third-instar hsp70-GAL4 larvae after heat shock. Photos adapted from Kennerdell and Carthew (2000). (See Color Plate 4 at the back of the book.)

mutants. In contrast, antisense or sense RNAs injected had an order-of-magnitude weaker interference activity than dsRNA (see Fig. 2). Another insect genetic model used for injection-induced RNAi is the red flour beetle, *T. castaneum*. The expression of the *Deformed* (*Dfd*) gene orthologues in *Tribolium* and *Drosophila* was successfully disrupted by microinjection of dsRNAs into embryos (Brown *et al.*, 1999), which proved that RNAi can be used for comparative analysis of gene functions in evolutionarily divergent organisms.

In addition to the genetic model insect species of *D. melanogaster* and *T. castaneum*, Hughes and Kaufman (2000) applied RNAi to a non-model insect species, the milkweed bug *Oncopeltus fasciatus*. They conducted dsRNA injection to dissect the functions of the Hox genes of *Dfd*, *proboscipedia* (*pb*) and *Sex combs reduced* (*Scr*) in mouthpart development. By analysing single and combined RNAi depletions, *Dfd*, *pb* and *Scr* were shown to function in determining the identity of the bug mouthparts; however, the exact roles of these Hox genes in milkweed bug are distinct from their counterparts' roles in *Drosophila* and *Tribolium* (Hughes and Kaufman, 2000).

In addition to insects, injection of dsRNA induces an RNAi response in a broad array of arthropods, including Arachnida and Crustacea (Aljamali *et al.*, 2003;

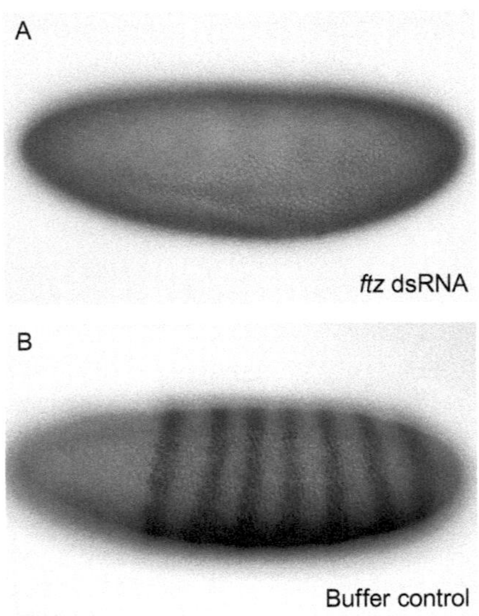

FIG. 2 Phenotype of *D. melanogaster* embryos injected with ds-*ftz* RNA. To determine whether dsRNA-mediated interference could occur in *Drosophila*, Kennerdell and Carthew chose the previously defined gene *fushi tarazu* (*ftz*) as a target. They synthesized dsRNAs *in vitro*, then injected dsRNAs into syncytial blastoderm embryos. Pictures are Ftz immunostaining of the stage 5 wild-type embryos injected with ds-*ftz* RNA and buffer control, respectively. Ftz expression disappeared from the dsRNA-injected embryo (A) but was detected in a normal seven-stripe pattern in the buffer-injected embryo (B). Photos adapted from Kennerdell and Carthew (1998).

Amparyup *et al.*, 2009; Aung *et al.*, 2011; Vargas-Vila *et al.*, 2010). The ixodid tick *Amblyomma americanum* is a vector of diseases such as human granulocytic ehrlichiosis, tularemia and Southern tick-associated rash illness. The tick contains several factors in its saliva, including proteins and prostaglandins, which modulate the host response at the tick-host interface (Aljamali *et al.*, 2002; Bowman *et al.*, 1997). Investigators constructed a dsRNA targeting the *A. americanum* gene encoding histamine binding protein (HBP). Female ticks injected with *HBP* dsRNA showed abnormal feeding pattern and reduction of histamine binding acticvity in isolated salivary glands, accompanying lower expression level of the *HBP* gene. These results demonstrate that RNAi can be employed as an important tool not only for studying the responsing mechamisms of salivary gland secreted proteins at the interaction between tick and its host, but also for controlling tick populations (Aljamali *et al.*, 2003).

To date, dsRNA injection as a powerful tool to assess gene functions has been extensively applied to a wide range of insect species (Isoe *et al.*, 2009; Jaubert-Possamai *et al.*, 2007; Liu *et al.*, 2010a,b; Maleszka *et al.*, 2007; Narasimhan *et al.*, 2004; Vallier *et al.*, 2009). Here, we present another case of RNAi in *Tribolium* to demonstrate its robust systemic RNAi response. *Tribolium* is an economically important agricultural pest species, which is exquisitely sensitive to dsRNA-mediated post-transcriptional gene silencing (Brown *et al.*, 1999; Tomoyasu and Denell, 2004; Tomoyasu *et al.*, 2008). Maternal RNAi in *Tribolium* was observed through injecting dsRNAs into a female pupa's haemocoel, resulting in an RNAi response in progeny (Bucher *et al.*, 2002). Tomoyasu and Denell (2004) used a green fluorescent protein (GFP) transgenic strain *pu11* with fluorescence in the wing and elytra discs to demonstrate that injection of *GFP* dsRNA into the larval body cavity inhibited *GFP* expression shortly after injection, and the RNAi effect continued throughout the pupal and adult stages. They used five different sizes of dsRNA (31, 69, 91, 256 and 520 bp), all caused loss of GFP expression in the larval, pupal and adult stages. Surprisingly, the silencing effect lasted at least 2 months, suggesting that some amplification and/or maintenance mechanism may exist in this beetle. According to their reports, the silencing effect was elicited even when as little as 5 pg of dsRNA per larva were injected. To determine whether the larval RNAi effect was local or systemic in *Tribolium*, dsRNA targeting *Tc-achaete-scutehomolog* (*Tc-ASH*) was injected, which revealed that larval RNAi could induce morphological defects in adult beetles, and also affected the entire body rather than being localized near the site of injection (Fig. 3).

3.2.4 *Feeding*

Delivering dsRNA via feeding has several advantages over injection. First, feeding causes little mechanical damage to insects. Further, feeding is convenient for the RNAi manipulation of a large number of individuals. One disadvantage of the feeding method is that it is difficult to measure or control

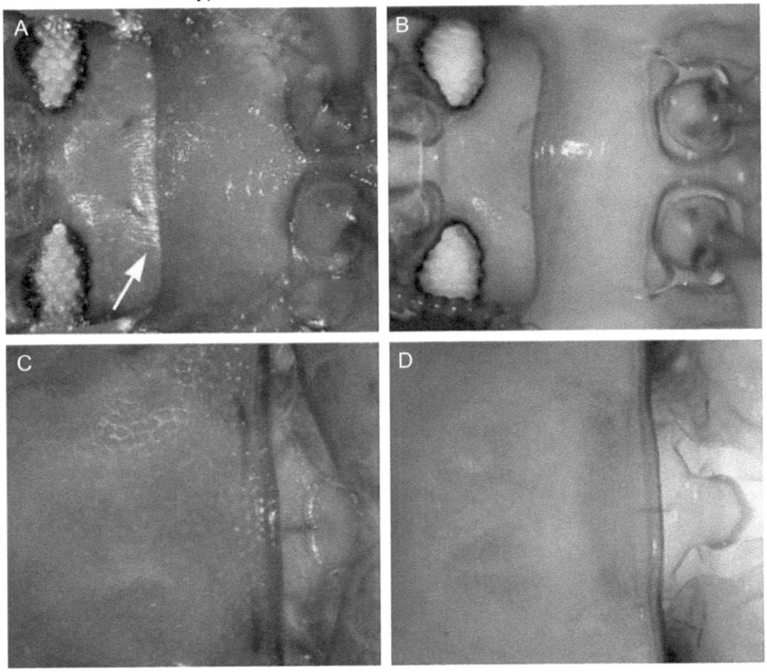

FIG. 3 *Tribolium castaneum* systemic larval RNAi revealed by *Tc-ASH* RNAi phenotypes. Non-injected *pu11* adult (A, C) and a *pu11* adult in which *Tc-ASH* dsRNA was injected during the last larval stage (B, D). Ventral view of the head and first thoracic sternum (A, B) and dorsal view of pro- and mesonotum (C, D). The entire wild-type adult cuticle is covered by small indentations, each associated with a sensory bristle (A, C). There is also a fringe of large bristles between the head and thorax (arrow in A). In *Tc-ASH* dsRNA-injected beetles, the cuticle was completely smooth (B, D) and had few sensory structures. The large bristles between head and thorax were also lost. (Photos reproduced courtesy of Yoshinori Tomoyasu and Robin E. Denell). (See Color Plate 5 at the back of the book.)

precisely the amount of dsRNA ingested (or absorbed from the gut). Although specific feeding habits of some insects restrict its range of applications, dsRNA feeding has been successful in a wide range of insect species. Recently, Huvenne and Smagghe (2010) summarized the results of RNAi in insects induced by dsRNA feeding. To date, this approach has been successful in at least 15 insect species belonging to seven different orders (Huvenne and Smagghe, 2010). Below, we divide insect RNAi through feeding into four classes, according to the methods of dsRNA production.

3.2.4.1. Ingestion of artificial diets supplied with dsRNAs As early as 1998, Fire *et al.* reported that when *C. elegans* worms were fed with bacteria expressing *GFP* dsRNA, the dsRNA was spread to the worms through ingestion and *GFP* expression was highly suppressed (Fire *et al.*, 1998). This finding indicated that dsRNAs can be transmitted from the environment to organisms. The ectoparasitic tick *Ixodes scapularis* transmits several pathogens, including *Borrelia burgdorferi* (a spirochete bacterium causing Lyme disease), to humans (Krause *et al.*, 2002). Soares *et al.* utilized capillary feeding of dsRNA to repress an anticomplement gene (*isac*) in *I. scapularis* nymphs; *isac* silencing resulted in dramatic tick weight decrease and reduced spirochete load. This research suggested that the oral delivery approach for specific gene silencing in *I. scapularis* nymphs could lead to ways to block transmission of *B. burgdorferi* (Soares *et al.*, 2005).

To down-regulate a carboxylesterase gene of light brown apple moth (*Epiphyas postvittana)*, dsRNAs directed against this gene were synthesized *in vitro*, and the third-instar larvae were fed on droplet solutions containing the dsRNA. Gene down-regulation was detected 2 days after the beginning of dsRNA feeding and maximal repression was achieved in 7 days (Turner *et al.*, 2006). To see if dsRNA could inhibit gene expression of salivary *nitrophorin 2* (*NP2*) of the triatomid bug (*Rhodnius prolixus*) and what impact this might have on apyrase activity in the saliva, the second-instar nymphs were fed with a solution containing 1 μg/μl of *NP2* dsRNA and about 42% reduction of the gene expression was then observed (Araujo *et al.*, 2006).

3.2.4.2. Ingestion of liposome-embedded or lipophilic siRNAs One modification of feeding protocols is to use lipid-encapsulated dsRNA rather than naked dsRNA. Liposomes have been used as nucleic acid transfection media for over 20 years; this approach originated from studies examining the ability of cationic lipids to deliver both DNA and RNA molecules into mouse and human cell lines (Felgner *et al.*, 1987; Malone *et al.*, 1989). Conjugation to lipophilic molecules (cholesterol, bile acids, and long-chain fatty acids) has been shown to increase siRNA uptake into cells and enhance gene silencing in mice (Wolfrum *et al.*, 2007). Efficient and selective uptake of these lipid-associated siRNAs depends on interactions with lipoprotein particles, lipoprotein receptors and transmembrane proteins. The efficacy of four commercially available transfection reagents inducing RNAi was evaluated in *D. melanogaster*. Larvae ingested dsRNA directed against the *gus* reported gene encapsulated in different kinds of liposomes for 2 h. Compared to a small (5–8%) reduction by naked dsRNAs, all liposomes facilitated some degree of RNAi effect in the isolated gut tissues (Whyard *et al.*, 2009). The authors suggested that Lipofectamine 2000 (Invitrogen) provided the highest RNAi efficacy with a 52% reduction in GUS activity. Quantitative RT-PCR confirmed that decreased GUS activity was closely correlated with the extent of *gus* transcript depletion (Whyard *et al.*, 2009). These data demonstrate that liposome-embedded dsRNAs can trigger a more efficient RNAi response in *D. melanogaster* larvae.

3.2.4.3. Ingestion of bacteria producing dsRNAs The dsRNAs incorporated into diets mentioned above were mostly synthesized *in vitro*. An alternative is to produce dsRNAs derived from insect genes in an engineered organism.

As introduced above, in *C. elegans* RNAi could be triggered by exposure to *E. coli* engineered to express gene-specific dsRNA (Timmons and Fire, 1998). Here, we take bacteria-mediated planarian RNAi (Newmark *et al.*, 2003) as an example to introduce this strategy. Fragments of 400–700bp derived from planarian target genes were inserted into the vector L4440 double T7 script II (Timmons *et al.*, 2001). The recombinant plasmids were introduced into competent HT115(DE3) cells that lacked RNase III and held a T7 polymerase that could simultaneously activate the transcription of sense and antisense strand in the presence of isopropyl β-D-thiogalactoside (IPTG). After IPTG induction, the cultured cells were centrifuged and resuspended in an artificial food mixture. Figure 4 shows the general strategy of bacteria-mediated RNAi. Using this method, a non-midgut chitin synthase gene (*CHSA*) was repressed in a lepidopteran pest, *S. exigua* (Tian *et al.*, 2009). *CHSA* is specifically expressed in ectodermal cells, including epidermal and tracheal cells. *CHSA* mRNA levels were substantially decreased compared to the negative controls at 7–9 days after larval ingestion of a high dose of bacteria expressing *dsCHSA*; however, there was no significant change in *CHSA* transcript levels in larvae fed a low dose of *dsCHSA* bacteria.

3.2.4.4. Ingestion of dsRNAs through feeding on transgenic plants The fact that RNAi can be triggered in both *C. elegans* and insects by ingestion of dsRNA-producing bacteria quickly led to the idea of engineering plants to express dsRNA, since a large proportion of insects are herbivores. As early as the 1980s, it was found that introducing the CP gene of TMV into tobacco enabled the plant to be resistant to TMV infection (Abel *et al.*, 1986). Subsequent studies suggested that in these virus-resistant plants, some of the transgenic mRNA molecules assumed the conformation of dsRNA, which

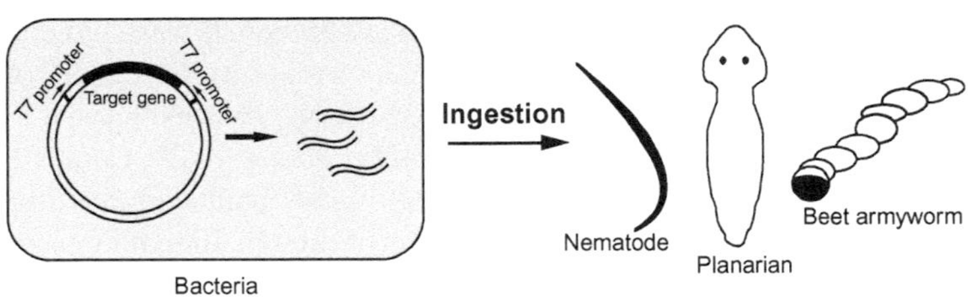

FIG. 4 The strategy of bacteria-mediated RNAi. The vectors producing dsRNAs in bacteria were constructed as follows. Fragments of target genes were inserted into the bidirectional T7 promoter vector pL4440 (Timmons *et al.*, 2001). The recombinant plasmids were introduced into HT115(DE3) competent cells. The dsRNAs were produced in the presence of isopropyl β-D-thiogalactoside (IPTG). After IPTG induction, cultured cells were collected and mixed with artificial food. RNAi was triggered when the animals ingested the diet containing dsRNA-expressing bacteria. Bacteria-mediated RNAi has been reported in *C. elegans*, a planarian and the lepidopteran insect *Spodoptera exigua* (Newmark *et al.*, 2003; Tian *et al.*, 2009; Timmons and Fire, 1998).

triggered sequence-specific RNA degradation in the cytoplasm (Waterhouse *et al.*, 1998). It was reported that plants expressing specific dsRNAs derived from pathogenic nematodes were resistant to nematode attack (Huang *et al.*, 2006). The dsRNA was designed against the root-knot nematode (RKN) parasitism-related gene *16D10*, And these RNAi signals were transmitted from plants to nematodes through worm ingestion of the plant materials, and then the RNAi response was triggered in worm bodies, resulting in reduced susceptibility of plants to RKN infection. However, earlier attempts at applying RNAi for pest control failed because the transgenic plants expressed insufficient dsRNA (Gordon and Waterhouse, 2007). Additional factors that affect the likehood of success using this approach are the sensitivity to RNAi, which varies across insects, and the silencing efficiency of different dsRNA fragments aimed at the same target gene, which is also highly variable. Our laboratory and one other group simultaneously reported progress in application of RNAi for generating insect-proof plants in 2007 (Baum *et al.*, 2007; Mao *et al.*, 2007). Both groups used transgenic plants to express dsRNAs designed to target insect pest genes and the genes were down-regulated after feeding the plant tissues. The details will be described in Section 5.

4 dsRNA uptake

A key step in using RNAi for pest control is the route of dsRNA uptake. A recent paper reviewed the mechanisms of dsRNA uptake in insects (Huvenne and Smagghe, 2010). There are two main pathways of dsRNA uptake in insect cells, one mediated by endocytosis and the other by transmembrane channels. Here, we summarize what is known about these uptake pathways in insects, nematodes (*C. elegans*) and plants (*Arabidopsis*).

4.1 ENDOCYTOSIS

In cultured *Drosophila* S2 cells, RNAi triggers are imported in an energy-dependent manner through a receptor-mediated endocytosis (Saleh *et al.*, 2006). In a screen based on preventing RNAi-induced lethality in S2 cells, several key components of RNAi in *Drosophila* were identified (Ulvila *et al.*, 2006). Besides Ago2, a known core component of the RNAi machinery, three more genes were for the first time found to participate in the RNAi response in *Drosophila*. One of these, *clathrin heavy chain* (*Chc*) was a component of the endocytic machinery, suggesting that endocytosis may play a role in dsRNA uptake by *Drosophila* cells. Blocking the endocytic pathway in S2 cells impaired RNAi, indicating that dsRNA fragments internalized by endocytosis are necessary for efficient gene silencing in S2 cells mediated by environmental triggers. In a further screen, the class C scavenger receptors SR-CI and Eater were shown to be combinatorially responsible for more than 90% of the internalized dsRNA in S2 cells. SR-CI is similar to the mammalian class A scavenger receptors and mediates uptake of bacteria as a phagocytic receptor in *Drosophila* S2 cells (Rämet *et al.*, 2001). Like SR-CI, Eater also mediates phagocytosis of bacterial pathogens in *Drosophila* (Kocks *et al.*, 2005). Homology was found with a scavenger receptor in the flesh fly (*Sarcophaga peregrina*) and *C. elegans*, but no mammalian homolog was found. When introduced into mammalian cells, SR-CI was able to mediate internalization of dsRNA molecules (Ulvila *et al.*, 2006).

Consistent with these results, several components associated with the endocytic pathway, including Chc, were identified in another investigation, which conducted a genome-wide screen for genes involved in dsRNA uptake using an 'RNAi of RNAi' approach (Saleh *et al.*, 2006). Together, these data suggest that endocytosis plays an important role in dsRNA internalization by *Drosophila* cells and dsRNA may be a kind of previously unrecognized ligands for these receptors. Endocytosis of dsRNA also functions in *C. elegans* because knockdown of components of the endocytic pathway by RNAi results in a 'loss-of-RNAi-function' phenotype in worms (Saleh *et al.*, 2006). Therefore, receptor-mediated endocytosis is likely to be a widespread mechanism for dsRNA uptake occurring across different classes of organisms (Price and Gatehouse, 2008).

4.2 TRANSMEMBRANE CHANNEL UPTAKE

In addition to dsRNA uptake into cells, the movement of dsRNA molecules between cells is also important in insect RNAi. Spread of the RNAi effect from the initial cells/tissues in which the dsRNA is applied or produced to other cells/ tissues where the silencing subsequently takes place results in systemic RNAi. Screening for genes critical for systemic RNAi resulted in the discovery of another important mechanism of dsRNA uptake. Injection of gene-specific dsRNA into one tissue of *C. elegans* could lead to the post-transcriptional silencing of that gene in other tissues, and the silencing effect was even transmitted to the worm's progeny. This phenomenon promoted identification of systemic RNA interference defective (sid) mutants of *C. elegans* by screening for worms resistant to systemic RNAi (Winston *et al.*, 2002). Three major complementation groups, *sid-1, sid-2* and *sid-3*, were isolated. *sid-1* mutants retain cell-autonomous RNAi but fail to show spreading of RNAi from one tissue to another. *sid-1* encodes a 776-amino acid protein with 11 transmembrane domains, and its homologues have been found in vertebrates, but not in *Drosophila*. Expression of *C. elegans sid-1* in *Drosophila* S2 cells enables energy-independent importing of dsRNA (Feinberg and Hunter, 2003). Shortly after this report, parental RNAi was disclosed in the flour beetle *T. castaneum*, as injection of dsRNA into the mother's haemocoel resulted in knockdown of zygotic genes in offspring embryos (Bucher *et al.*, 2002). With the increasing release of insect genomes, we know that almost all species of insects possess *SID-1* orthologues except those of Diptera (Honeybee Genome Sequencing Consortium, 2006; Tribolium Genome Sequencing Consortium, 2008; Xu and

Han, 2008). Some species may have several copies; for example, there are three orthologues in *T. castaneum* (Tomoyasu *et al.*, 2008) and *Bombyx mori* (Honeybee Genome Sequencing Consortium, 2006), and at least two in the cotton bollworm *H. armigera* (our unpublished data). Recent studies of *C. elegans* showed that SID-1 is not selective for dsRNA length when functioning as a dsRNA transporter (Shih *et al.*, 2009). Based on the analyses of systemic RNAi in *Tribolium*, *SID-1* orthologues in insects were presumed to have a function similar to that in *C. elegans*. In contrast to the expectation that *sid* homologues were required for RNAi in *Tribolium*, silencing of the three *sid-1* orthologues of *T. castaneum*, either individually or all together, had no effect on RNAi, suggesting that *sid* genes may be not required for systemic RNAi in *Tribolium*. However, involvement could not be excluded because multiple dsRNA triggers could weaken the efficiency of RNAi in *C. elegans* and *Tribolium* (Parrish *et al.*, 2000; Tomoyasu *et al.*, 2008), as possible competition for the cell-autonomous core RNA machinery between three dsRNAs resulted in incomplete knockdown of the *sid* genes (Tomoyasu *et al.*, 2008). At present, whether insect *SID-1* homologues are involved in systemic RNAi is an open question, as there is no solid evidence that the absence of *SID-1* homologues in insects results in failure of systemic RNAi. Interestingly, insects do not have homologues of SID-2, which is required for environmental RNAi in *C. elegans*. Environmental RNAi refers to the phenomenon that silencing triggering molecules are taken up from the environment (Winston *et al.*, 2007).

4.3 OTHER COMPONENTS IN RNAI SIGNAL TRANSPORT

Although *C. elegans* tissues require SID-1 to import silencing signals, export of RNA silencing from *C. elegans* tissues does not, which means that there is an unidentified mechanism mediating export of the RNAi triggers (Jose *et al.*, 2009). Distinct mechanisms to transport RNAi triggers between cells have been identified in plants and animals. In plants, transport of RNAi triggers between adjacent cells is thought to occur through intercellular bridges called plasmodesmata (Dunoyer *et al.*, 2005), and long distance RNAi transmission occurs via the phloem vascular tissue (Crete *et al.*, 2001; Voinnet *et al.*, 1998). In a genetic *silencing-movement-deficient* (*smd*) screen, components participating in intercellular movement of silencing signals in *Arabidopsis* were identified. *SMD1* and *SMD2*, allelic to *RDR2* (encoding an RdRP) and *NRPD1a* (encoding a subunit of a plant-specific DNA-dependent RNA polymerase IV), are essential for intercellular but not intracellular RNAi (Dunoyer *et al.*, 2005, 2007). *NRPD1a*, *RDR2* and *DCL3* are required for reception of long distance RNAi in the shoot (Brosnan *et al.*, 2007). An SNF2 domain-containing protein, CLASSY1, acts together with RDR2 and NRPD1a in the nucleus in spreading of transgene silencing (Smith *et al.*, 2007).

It was reported that vesicles (exosomes) containing mRNAs and miRNAs could be delivered into mammalian cells (Valadi *et al.*, 2007). Participation of vesicle-related proteins in systemic RNAi was discovered in worms. Three proteins, RSD-2, RSD-3 and RSD-6, were identified as essential for systemic RNAi in the *C. elegans* germ line (Tijsterman *et al.*, 2004). RSD-3 contains an epsin N-terminal homology (ENTH) domain which is involved in vesicle trafficking. This result is supported by a subsequent report that knockdown of some vesicle transport protein genes resulted in resistance to feeding RNAi (Saleh *et al.*, 2006). Therefore, ingested RNAi triggers can be loaded into vesicles. A possibility is that environmental RNAi triggers are imported into cells through SID-1 channel and/or scavenger receptors. A subset of these internalized dsRNAs are enveloped into vesicles, and only vesicle-embedded dsRNAs are then transmitted to adjacent cells or cell cavities. The mechanism for such transport has not yet been identified.

Systemic RNAi in plants and worms is typically attributed to RdRPs, because RdRPs are assumed to be key proteins for RNAi signal amplification. Surprisingly, RdRP-deficient *Tribolium* exhibited robust systemic RNAi (Tomoyasu and Denell, 2004; Tomoyasu *et al.*, 2008). Although the homologues of canonical RdRPs are absent in insect genomes (Gordon and Waterhouse, 2007), a subunit of the RNA polymerase II with RdRP activity has been identified to be involved in RNAi and transposon suppression in *D. melanogaster* (Lipardi and Paterson, 2009). Thus, insects may possess a distinct mechanism for spreading RNAi. Understanding the process of systemic RNAi in insects would bring significant progress to the development of RNAi-based approaches to pest control, as well as to functional genomics research.

5 Insect pest control based on RNAi

Since its discovery 13 years ago, RNAi has evolved from a research tool into a potentially powerful technology with applications in human medical therapy, nutrition quality improvement, plant disease/pathogen control and pest control. The reports that dsRNAs designed according to insect gene sequences could be delivered via transgenic plants (Baum *et al.*, 2007; Mao *et al.*, 2007) presented the prospect for a new generation of pest control technology.

5.1 PEST GENES FOR TARGETING

Probably the first step in development of RNAi-based technology of plant protection is the selection of effective and specific target genes in the targeted pests. In the life cycle of a given insect, each set of genes is critical for certain developmental aspects, such as growth, metamorphosis, food digestion, detoxification and so on. The silencing or overexpression of these genes would result in altered insect growth and development, or even in lethality. Housekeeping genes are one set of candidate targets. When designing RNAi strategy for enhancing corn resistance to WCR, genes encoding proteins with essential functions in WCR were considered the best RNAi targets for causing lethality. In a feeding assay to assess WCR sensitivity to ingested dsRNAs, many of the target genes whose suppression resulted in larval stunting and/or mortality were housekeeping genes, including those encoding actin, α/β-tubulin, ATP synthase chain A, ribosomal protein and ADP/ATP translocase (Baum *et al.*, 2007).

Detoxification or defence-related genes are also of special value as pest control targets. They offer a potential advantage over housekeeping genes in that they are usually less conserved, thus increasing the silencing specificity and reducing the off-target side effects. Plants produce phytoalexins to resist pathogenic microbes and herbivores that would attempt to consume them. For example, cotton plants accumulate gossypol and related sesquiterpene aldehydes (Luo *et al.*, 2001; Tan *et al.*, 2000), which are toxic to many organisms (Du *et al.*, 2004; Stipanovic *et al.*, 2006). To counteract plant defences, insects such as the cotton bollworm have evolved adaptive mechanisms which involve a set of genes whose products metabolize the secondary metabolites accumulated in plant tissues (Mazumdar-Leighton and Broadway, 2001; Wittstock *et al.*, 2004). Despite the adverse effect of gossypol, cotton bollworm larvae prefer to feed and complete their life cycle on cotton plants, implying that bollworms have mechanisms that enable them to either detoxify or tolerate gossypol. The cotton bollworm is one of the most devastating agricultural pests, causing severe yield loss in both cotton and other crop species. Our group is interested in elucidation of the gossypol biosynthesis pathway and the mechanisms of cotton bollworm tolerance to gossypol. In order to enrich genes potentially involved in gossypol response, we constructed a cDNA library from the midgut of fifth instar larvae after a 1-day administration of gossypol ($1\,mg\,g^{-1}$) added to the artificial food. A cDNA array analysis of midgut gene expressions identified several gossypol-inducible genes, among which one contained a 1581-bp open reading frame encoding a P450 monooxygenase, CYP6AE14. This gossypol-inducible P450 (GIP) belongs to the CYP6 family, with 67% protein sequence identity to CYP6AE12 of *H. armigera*. In insects, CYP6 enzymes have been shown to contribute to insect tolerance to xanthotoxins (Scott *et al.*, 1998), and *CYP6AE1* is furanocoumarin-inducible (Li *et al.*, 2004). Since induced metabolism of xenochemicals is a common feature of insect detoxification systems (Berge *et al.*, 1998; Scott and Wen, 2001), this gossypol-inducible P450 is likely a key component of *H. armigera* defence response to gossypol. Since larvae required an elevated level of *CYP6AE14* expression to feast on food supplemented with gossypol, we predicted that suppression of *CYP6AE14* would weaken cotton bollworm tolerance to gossypol (Mao *et al.*, 2007).

Genes closely related to insect developmental patterns are another class of targets. Silencing the *Bursicon* gene of the silkworm by dsRNA injection resulted in inhibition of wing expansion (Huang *et al.*, 2007), and disturbance of other developmental genes of silkworm was lethal (Dai *et al.*, 2008; Liu *et al.*, 2010a; Xiang *et al.*, 2011). In *Tribolium*, RNAi down-regulation of a number of hormone-related genes, such as *ecdysone receptor* (*EcR*), *Ultraspiracle* (*USP*), *Ecdysis-triggering hormone* (*ETH*) and *eclosion hormone* (*EH*), has proved to be very effective and significantly affected beetle development (Arakane *et al.*, 2008; Tan and Palli, 2008). In our experiments, not all selected target genes were down-regulated by dsRNA (our unpublished data), which is consistent with the results obtained from experiments of WCR RNAi, which demonstrated that 125 of the 290 dsRNA species created significant larval mortality and/or stunting (Baum *et al.*, 2007). These data suggested that even within a species, RNAi response is diverse across individual genes.

The length of dsRNA is another important factor for dsRNA design. Longer dsRNA is preferentially imported by *Drosophila* S2 cells overexpressing the *C. elegans sid-1* gene (Feinberg and Hunter, 2003). *In vivo*, injection of a 50-bp dsRNA into intestinal cells did not induce systemic RNAi in *C. elegans*, suggesting that long initial dsRNAs rather than siRNAs are critical for the systemic RNAi response (Feinberg and Hunter, 2003). Moreover, for *Drosophila* S2 cells, RNAi was dependent on the length of the dsRNA. A mixed pool of 21-bp siRNAs cleaved from a 1000-bp dsRNA fragment by RNase Dicer failed to enter S2 cells, even though it effectively provoked gene silencing after transfection (Saleh *et al.*, 2006). In a dsRNA soaking experiment, longer dsRNAs administered more potent RNAi than shorter dsRNA (Saleh *et al.*, 2006). The length of dsRNA for transgenic plants in our experiments was 469 and 101 bp for cotton bollworm *dsCYP6AE14* and *dsGSTI*, respectively (Mao *et al.*, 2007). Huvenne and Smagghe (2010) reviewed dsRNA length in studies using feeding RNAi: it was in the range of 134–1842 bp except one example in which 21 bp siRNAs reduced expression of the acetylcholine esterase gene in *H. armigera*. Thus, to achieve high efficiency of plant-mediated RNAi in insects, screening targets is a prerequisite, and an appropriate length of dsRNA also should be identified.

5.2 PLANT-MEDIATED INSECT RNAI

To suppress *CYP6AE14* expression in the midgut of cotton bollworms, a hairpin structure with an inverted repeat derived from the *CYP6AE14* coding sequence was constructed and transferred into *Arabidopsis* and *Nicotiana tabacum* (tobacco). Third-instar larvae of *H. armigera* were transferred from artificial diets to leaves of the transgenic plants. Northern blot analysis showed the presence of small RNA fragments of *CYP6AE14* in midgut 2 days after transfer (Fig. 5A). Four days after transfer, *CYP6AE14* transcript and protein levels were down-regulated in midgut of the larvae reared on *dsCYP6AE14* leaves compared to those kept on *dsGFP* or wild-type leaves (Fig. 5B and C). Importantly, we found that transgenic plants with higher dsRNA expression levels had stronger silencing effects on *CYP6AE14* expression. Furthermore, larval growth was slightly retarded by the *dsCYP6AE14*, but not by the *dsGFP* or the wild-type plants (Fig. 5D and E). These findings suggested that the higher the level of dsRNA expression in plants was, the more efficient the RNAi response in insects

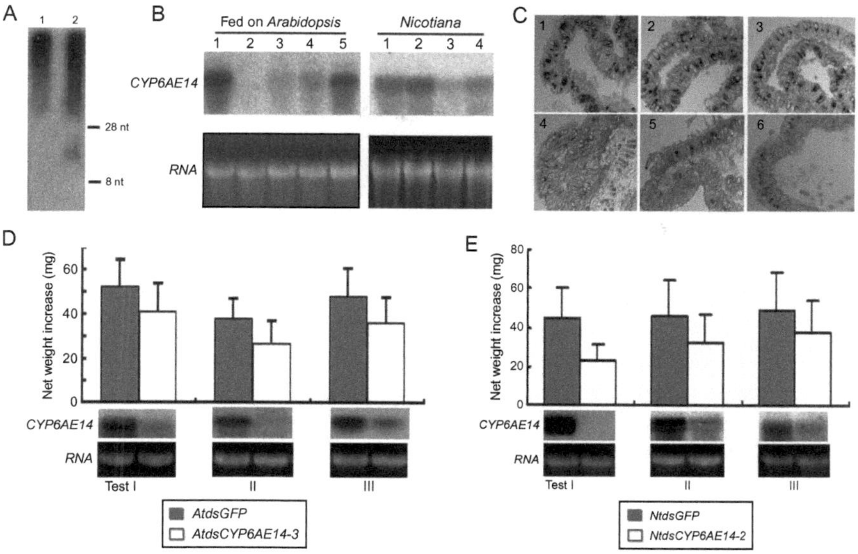

FIG. 5 *CYP6AE14* suppression in cotton bollworm larvae fed on transgenic plant leaves expressing *CYP6AE14* dsRNA. (A) Northern blot of the small RNAs of *CYP6AE14* in midgut of bollworm (*Helicoverpa armigera*) larvae fed on *AtdsGFP* (1) and *AtdsCY-P6AE14-3* (2) leaves for 2 days. Small RNAs of *CYP6AE14* was only detected in larvae fed on *dsCYP6AE14* plants (lane 2), not those on control plants (lane 1). (B) Northern blot of *CYP6AE14* transcripts in midgut of larvae fed on leaves of different transgenic plants for 4 days. *Arabidopsis thaliana* leaves were taken from *AtdsGFP* (1), *AtdsCY-P6AE14-2* (2), *AtdsCYP6AE14-3* (3), *AtdsCYP6AE14-4* (4) and *AtdsCYP6AE14-5* (5); and *Nicotiana tabacum* leaves from wild type (1), *NtdsGFP* (2), *NtdsCYP6AE14-2* (3) and *NtdsCYP6AE14-4* (4) plants, respectively. The *CYP6AE14* transcript levels were significantly down-regulated in midguts of larvae fed on *dsCYP6AE14* plants. (C) Immunohistochemical detection of CYP6AE14 proteins in the midgut of larvae fed on *NtdsGFP* (1–3) and *NtdsCYP6AE14-2* (4–6) leaves for 4 days. (D and E) Average weight increase (upper) and midgut *CYP6AE14* transcript level (lower) of larvae fed on *AtdsGFP* and *AtdsCYP6AE14-3*, or *NtdsGFP* and *NtdsCYP6AE14-2* leaves for 4 days. Photos adapted from Mao *et al.* (2007).

(Mao *et al.*, 2007). To determine whether plant-mediated insect RNAi, exemplified here with *CYP6AE14*, was applicable to other genes expressed in the midgut, we selected another *H. armigera* gene, *GST1* (GenBank accession no. EF033109), which encodes a glutathione-S-transferase. Four days after the third-instar larvae were transferred to leaves of the dsRNA-producing *Arabidopsis* plants (*AtdsGST1*), levels of *GST1* transcripts in the midgut were reduced.

Both long dsRNAs and short interfering RNAs of *CYP6AE14* were detected in the transgenic plants. It was then interesting to address which form was the main or more efficient one in insect gene silencing. To generate plants that predominately produce intact, long dsRNAs of the transgene, we employed an *Arabidopsis* triple mutant lacking three of the four *Arabidopsis* Dicer-like nucleases (Xie *et al.*, 2004).

The triple *DCL* mutant plants expressing the *dsCYP6AE14* exhibited a dramatic increase in long dsRNAs and a more pronounced effect in eliciting *CYP6AE14* silencing in the bollworm midgut. A plausible explanation is that long dsRNAs are the main form transmitted into midgut cells, which is consistent with the results obtained with *Drosophila* S2 cells, where RNAi was dependent on the length of the dsRNA (Saleh *et al.*, 2006).

Following the encouraging results obtained with *Arabidopsis* and tobacco, we generated transgenic *dsCYP6AE14* cotton plants. Bollworms reared on T2 cotton plants exhibited drastically retarded growth compared to those on control cotton plants. Importantly, the transgenic plants were less damaged by bollworms than were the non-transgenic control (Fig. 6A). Western blot assay showed that CYP6AE14 protein levels were obviously lower in the larvae fed on transgenic cotton plants for 3 days (Mao *et al.*, 2010).

The WCR is one of the most devastating corn pest species in America. After hatching, WCR larvae begin chewing root hairs and small roots (http://www.ext.colostate.edu/PUBS/insect/05570.html). To protect corn against WCR infestation, a group of scientists took an approach parallel to ours. Maize plants were engineered to express dsRNAs targeting a vacuolar ATPase (V-ATPase) gene of WCR (Baum *et al.*, 2007). To test the feasibility of the RNAi-based strategy of WCR control, genes whose disturbance was predicted to cause insect lethality were selected as targets. In the first round assay, *in vitro* synthesized dsRNAs were laid on the surface of WCR artificial diet at concentrations ranging from 520 to 780ng/cm^2. Further feeding assays found that dsRNA species directed against genes encoding V-ATPase subunits and α-tubulin also worked well even below 52ng/cm^2. To engineer corn plants with enhanced resistance to WCR, a fragment of the putative V-ATPase A coding region of WCR was placed in a transformation cassette under the CaMVe35S promoter, which resulted in the production of the dsRNAs in transgenic plants. The effectiveness of this pest control approach was tested by assessing the damage induced by rootworm feeding on F1 hybrid plants. As shown in Fig. 6B, the transgenic maize plants showed a remarkable reduction in feeding damage by WCR compared to the controls, after a 3-week period of feeding (Baum *et al.*, 2007). This, together with the example of our own, proved that transgenic plants expressing dsRNAs targeting insect genes could elicit insect RNAi and protect themselves from herbivorous pest damage.

5.3 OTHER FEASIBLE METHODS

5.3.1 *dsRNA expression in chloroplast*

Since small RNAs can be produced in bacteria and the plastid genome is of bacterial origin, it is possible to engineer chloroplasts to produce dsRNAs. Expressing foreign genes in chloroplast offers several advantages over nuclear expression. First, chloroplast transformation may result in high expression levels due to numerous copies of chloroplasts in a cell. Taking the Bt crystal

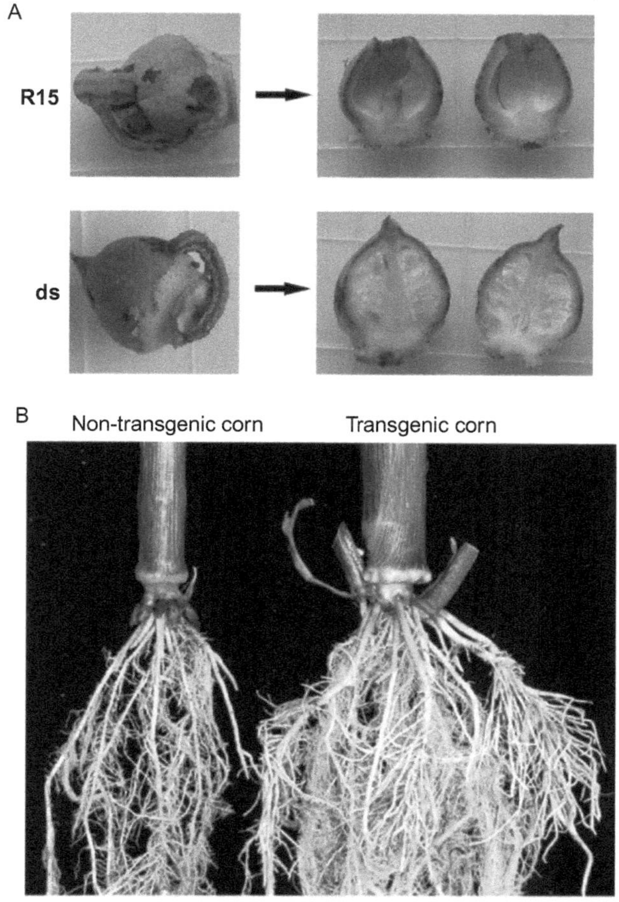

FIG. 6 dsRNA transgenic plants were less damaged. (A) *dsCYP6AE14* transgenic cotton plants were less damaged by bollworms. To test the protection of cotton bolls by dsRNA transgenic cotton, *Helicoverpa armigera* larvae that had previously been reared on leaves of the wild type and *dsCYP6AE14* plants for 10 day were transferred to bolls (about 9 days post-anthesis). After 1 day, the larvae from wild type leaves burrowed into bolls and consumed the contents (upper panel), whereas the larvae from *dsCYP6AE14* leaves chewed only shallow gouges in the boll surface, and the boll contents were almost intact (lower panel). These data indicate that the transgenic cotton plants expressing the insect P450 dsRNA were less damaged by cotton bollworms in comparison with the untransformed control. Photos adapted from (Mao *et al.*, 2011). (B) F1 corn plants expressing a V-ATPase A dsRNA were protected from WCR feeding damage. To test whether corn plants expressing a WCR-derived dsRNA were protected from rootworm feeding damage, Baum *et al.* (2007) transformed a V-ATPase A dsRNA expression cassette into corn plants. The maize transformants were backcrossed with an inbred line to generate F1 hybrid progeny. F1 hybrid plants were evaluated using a growth chamber assay to assess rootworm feeding damage. The plant on left is a non-transgenic control with average root damage, whereas the dsRNA transgenic plant on the right shows obvious root protection. (Photos reproduced courtesy of James Baum and James Roberts). (See Color Plate 6 at the back of the book.)

insecticide as an example, it was reported that amplification of an unmodified CrylAc coding sequence in chloroplasts up to approximately 10,000 copies per cell resulted in the accumulation of protoxin to a level of 3–5% of soluble proteins in tobacco leaves. The plants showed an extremely high insecticidal activity to larvae of *Heliothis virescens*, *H. zea*, and *S. exigua* (McBride *et al.*, 1995). Secondly, traits encoded by chloroplast are predominantly maternally inherited in most plants, so that the transgene is less likely to be transmitted to non-transgenic plants. With the achievement of stable chloroplast genome transformation in green plants except tobacco (Dufourmantel *et al.*, 2004; Hou *et al.*, 2003; Kumar *et al.*, 2004; Liu *et al.*, 2007; Okumura *et al.*, 2006; Ruf *et al.*, 2001; Sidorov *et al.*, 1999), chloroplast transformation shows a great potential for crop protection against pests.

5.3.2 *dsRNA production in natural symbiotic bacteria*

According to a recent estimation, as many as 20% of insects are obligately associated with symbiotic microorganisms; insects have the capacity to use microbial products, and the interactions have greatly promoted their co-evolution (Feldhaar and Gross, 2009). One such example is the pea aphid (*Acyrthosiphon pisum*) and its symbiotic bacterium *Buchnera aphidicola* (*Buchnera* APS). The pea aphid lacks the capacity to synthesize arginine, which is produced by *Buchnera* APS (Wilson *et al.*, 2010). This relationship provides an opportunity to engineer *Buchnera* APS to produce dsRNAs targeting critical genes of its host, or a protein that is effective against this host.

5.3.3 *dsRNA production via the VIGS system*

Introducing viral genes into plants could result in transgenic plants resistant to the virus even when no transgene protein is translated (Lindbo and Dougherty, 1992). Later, it was shown that sequence-specific RNA degradation was the factor behind this virus-resistance (Waterhouse *et al.*, 1998). Based on this principle, investigators exploited the virus-induced gene silencing (VIGS) system based on the antiviral defence mechanism of plants as a powerful tool for plant reverse genetics. Vectors containing genome sequence of a virus combined with host genes were inoculated into plants which then induced loss-of-function of corresponding genes through production of dsRNA (Purkayastha and Dasgupta, 2009). It may be possible to imitate the operation of VIGS in plants by replacing the host with pest genes.

5.3.4 *Co-expression of "effector" proteins with dsRNAs*

It can be a common situation that insect RNAi triggered by ingestion does not silence the target gene completely. Enhancing RNAi efficiency thus may improve the success of this approach. In the insect midgut, the peritrophic membrane (PM) is the first line of defence. PM is composed of chitin fibrils and proteins, and most PM proteins contain chitin-binding domains, which are characterized by conserved cysteine residues and aromatic amino acids. Chitinase is expressed in the midgut only during moulting, when chitin of the PM is presumably degraded. Thus, it was supposed that, when insects feed on plants, a constitutively expressed chitinase may disturb normal development. To test this, larvae of the budworm *H. virescens* were placed on transgenic tobacco plants expressing a tobacco hornworm (*Manduca sexta*) chitinase gene. Growth of the budworm larvae and feeding damage to plants were significantly reduced (Ding *et al.*, 1998). Because they damage the PM, protease and chitinase could be potential partners of dsRNAs.

Snowdrop lectin (*Galanthus nivalis* agglutinin; GNA) has an ability to bind to the gut epithelium, and can be detected in the haemolymph of insects which ingested the plant tissues (Fitches *et al.*, 2001). This suggests a possibility that GNA may be used as a carrier to deliver fused proteins to the haemolymph of target species. There are concerns about possible consequences to vertebrates ingesting snowdrop lectin. An alternative is to find an insect-specific lectin with no adverse effects on vertebrates. In addition, it was reported that efficient siRNA delivery into mammalian primary cells was achieved using a peptide transduction domain-dsRNA binding domain (PTD-DRBD) fusion protein. DRBDs bind to siRNAs and allow PTD-mediated cellular uptake (Eguchi *et al.*, 2009). For insect siRNA delivery, it may be feasible to replace PTD with a lectin to create lectin-DRBD fusion proteins; the lectin fusion carrying dsRNAs bound by DRBD may enhance dsRNA uptake across the gut. An encouraging development on this front is the isolation of a mannose-binding lectin from yam (*Dioscorea batatas*, synonym *Dioscorea polystachya*) tubers, which strongly bound to larval brush border and PM cells, resulting in insecticidal properties against *H. armigera* (Ohizumi *et al.*, 2009). Although no evidence showed that yam lectin could penetrate insect cell membrane like GNA, we speculate that yam lectin fusion proteins could contribute to pest control in the future.

6 Discussions

Since its discovery, RNAi has been considered one of the most powerful tools in many areas of medicine, agriculture and biological research. One of the promising areas of RNAi application is to develop a new generation of insect-proof plants.

Insects form the most diversified group of organisms with more than 1 million species, and different insect taxa show diverse degrees of RNAi responses. For example, the beetle *T. castaneum* is highly sensitive to RNAi injection, with effective dosages as low as 5 pg/larva (Tomoyasu and Denell, 2004), whereas at least 20 μg of dsRNA must be injected into lepidopteran larvae to obtain gene down-regulation (Huang *et al.*, 2007). A very recent article collected the published and unpublished data from the studies about RNA interference in Lepidoptera and clearly defined the insect sensitivity to RNAi as the amount of dsRNA required to execute silencing per mg tissue. Because the sensitivity varies in different species, as well as in different tissues of the same organism, the authors recommended that researchers use the concentrations of dsRNA (as opposed to amounts of dsRNA), such as μg (RNA)/mg (body) in the description of experiments (Terenius *et al.*, 2011). The public database http://insectacentral.org/RNAi aims to continue to gather information will further our understanding of RNAi in Lepidoptera.

RNAi works well in the family of Saturniidae, but some species (*Bicyclus anynana*, *Chrysodeixis includens* and *Spodoptera littoralis*) have been so far proven resistant to application of dsRNA. A comprehensive survey of literatures suggested that RNAi is usually successful for genes involved in immunity and defence responses expressed in hemocytes, fat body and midgut, but often failed in genes expressed in epidermal tissues (Terenius *et al.*, 2011). On the other hand, insects at different developmental stages have various responses to RNAi. For silkworm, the embryo and pupal stages are more sensitive to RNAi than the larval and adult stages (Pan *et al.*, 2009; Xiang *et al.*, 2008). *D. melanogaster* is relatively insensitive to environmental and systemic RNAi, but overexpression of dsRNAs within cells readily triggered RNAi in tissues resistant to systemic RNAi (Miller *et al.*, 2008). This suggests that the diversity of RNAi sensitivity in different species is not due to the RNAi core machinery, but rather to the abilities of penetration and transmission of the RNAi signal through cells and tissues (Belles, 2010; Terenius *et al.*, 2011). Degradation mechanisms that are able to remove alien RNA may also contribute to differences in sensitivity. In the case of differences in sensitivity between tissues and stages of the same organism, levels of components related to dsRNAs uptake and intracellular RNAi may regulate these differences. For example, RNAi down-regulation of salivary gland genes in the mosquito *A. gambiae* required high doses of dsRNA, which is consistent with the lower abundance of transcripts of *Dicer* and *Argonaute* genes in salivary gland than in other tissues (Boisson *et al.*, 2006). Another factor that could influence the potency of RNAi is the regulatory pattern of target genes. For example, genes with efficient feedback regulation network might readily counteract depletion of mRNA levels through higher rates of transcription (Belles, 2010).

Besides differences in inherent sensitivity to RNAi and target gene regulatory mechanisms, an efficient delivery pathway holds the key to the application of the RNAi-based pest control. Despite the exciting progress in engineering plants to trigger RNAi in herbivorous insects, many questions remain open. For example, it is interesting to examine if the plant-induced gene silencing effect can be amplified and spread among insect tissues. Ingested dsRNAs can trigger systemic RNAi in *C. elegans* because the nematodes possess RdRP and transmembrane SID-1 proteins for RNAi signal amplification and transmission, respectively (Feinberg and Hunter, 2003; Sijen *et al.*, 2001, 2007; Winston *et al.*, 2002). However, cotton bollworm and WCR, the two insect species reported so far to be sensitive to plant-mediated RNAi, have transporters homologous to SID-1 but lack RdRPs (Gordon and Waterhouse, 2007). Although SID-1 takes charge only of dsRNA import, not in export, and there is no compellent evidence supporting the idea that *sid* gene are necessary for systemic RNAi in *Tribolium*; however, a role for *sid* in insect systemic RNAi also cannot be conclusively ruled out at present. Moreover, RdRP activity of a subunit of the RNA polymerase II had been documented in *D. melanogaster* (Lipardi and Paterson, 2009). Therefore, we suggest that uptake of dsRNA from the gut is probably mediated by transporters such as SID-1 proteins, or by scavenger receptors associated with endocytosis (see Fig. 7). Then small RNAs absorbed directly from guts or generated by insect Dicers from long dsRNAs induce gene silence in insect cells. More work is needed to determine the dominant manner (endocytosis, transmembrane channel or other) by which midgut cells take up intestinal dsRNAs. Further understanding of the complexities of insect RNAi amplification and transmission would facilitate the targeting of specific insect pests.

Because the aim of pest control is to decrease the pest population, an important requirement is to limit the side effects of the RNAi. One of the concerns about RNAi technology is that unwanted gene products with sequence similarity to target gene may also be down-regulated. There are two types of off-target effects:

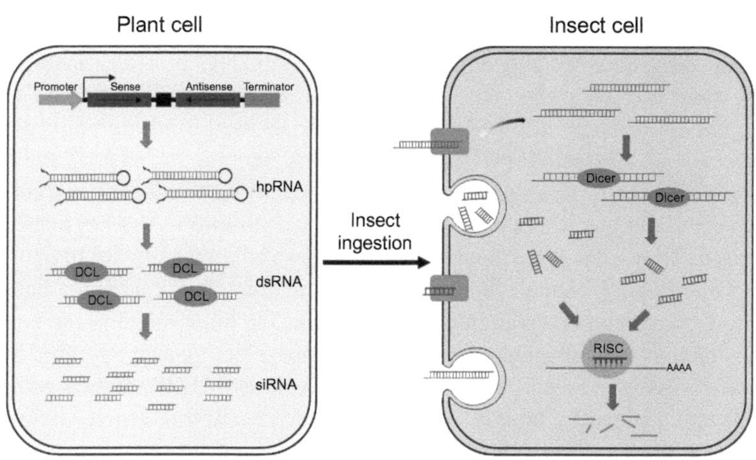

FIG. 7 Insect RNAi mediated by transgenic plants. Transgenic plants express hairpin RNAs (hpRNA) directed against an insect gene. HpRNAs are diced into the intermediate product of dsRNAs, and then processed into siRNAs by plant Dicer enzymes. The dsRNAs and/or the siRNAs are orally delivered into larvae feeding on transgenic plants by SID-1 homologue-mediated transmembrane channel, scavenger receptor-mediated endocytosis or other pathways. The siRNAs, ingested directly from the plants or processed by insect Dicers after ingestion of long dsRNAs, elicit cleavage of the target mRNA in RISC in midgut cells. (See Color Plate 7 at the back of the book.)

different genes within the species and cross-species. In normal RNAi application in functional research, only the off-target effects within a species should be avoided. Through analysing the occurrence of off-target effects in *Drosophila*, it has been demonstrated that most genes in the *D. melanogaster* genome contain two or more short (21 bp) sequences that show sequence similarity (containing three or fewer mismatches) to the same off-target gene (Seinen *et al.*, 2010). To avoid off-target effects in *D. melanogaster*, a bioinformatics approach was developed to design dsRNAs that do not contain predicted off-targets in *Drosophila* (Seinen *et al.*, 2010).

Regarding to the RNAi-based pest control, the major concern is to avoid cross-species off-target effects. Here, we cite two cases to illustrate this kind of off-target effects and the corresponding solutions. The dsRNAs targeting β-tubulin, V-ATPase A and E showed a strong RNAi response with high larval mortality in WCR (Baum *et al.*, 2007). These dsRNAs were also used to test two other coleopteran plant pests: southern corn rootworm (SCR; *Diabrotica undecimpunctata howardii*) and Colorado potato beetle (CPB; *Leptinotarsa decemlineata*). The dsRNA feeding led to significant larval mortality in SCR and CPB, although higher doses were required than those used for WCR. WCR and CPB share only 83% sequence identity in V-ATPase A target region and 79% in V-ATPase E, respectively. However, the cross-silencing was successfully escaped by Wong and his colleagues when they silenced *γ-tubulin* in four *Drosophila* species (Whyard *et al.*, 2009). The *γTub23C* gene is highly conserved in four *Drosophila* species (*D. melanogaster*, *D. sechellia*, *D. yakuba*, and *D. pseudoobscura*), with 79–96% sequence identity in coding regions between each species. The untranslated region (UTR) is often unconserved among homologous genes. The sequence divergence of 3′UTR of *γTub23C* among the four species allowed these regions to be used for designing dsRNAs. No species was affected by *γTub23C* dsRNA specific to any of the other *Drosophila* species. This suggests that RNAi can be carried out to target only one species as long as there is sufficient sequence variation. Another question is whether dsRNA transgenic plants could threaten vertebrates owing to off-target effects of RNAi. Although there is no report of this phenomenon to date, the potential risk should be taken into consideration. We think that this can be ruled out by detailed analysis of gene sequences before selecting a target gene, and by the use of less conserved gene and regions (such as 3′UTR). Hence, although housekeeping genes are probably more potent as pest control targets, use of non-housekeeping genes may decrease off-target risks, since they are often less conserved. This is one of the reasons that, in our experiments, we selected a defence-related P450 gene (*CYP6AE14*) for silencing (Mao *et al.*, 2007). Although we would not exclude the off-target possibility due to absence of complete genome information of *H. armigera*, high sequence specificity of RNAi effect can be maximized with a careful bioinformatic selection of target genes. Therefore, genome data accumulation will serve not only efficient pest control, but also environmental safety.

RNAi as an innovative, promising strategy for insect pest control has been explored in the past few years. All the same, there are additional challenges that must be addressed before the strategy can be broadly deployed. Take producing transgenic crop plants as an example: dsRNA overexpression via CaMV 35S promoter may suffer from transgene silencing, especially the siRNA induced transcriptional gene silencing through DNA methylation. Alternatives include employing inducible promoter (e.g. wounding inducible promoter), expressing dsRNA in plasmids (chloroplast), or expression via the VIGS system. Another problem is that plant-mediated insect RNAi does not kill the pests immediately like Bt; rather, it protects the crops by decreasing the damage. However, it also can be an advantage from ecological points of view. Perhaps this will decrease the selective pressure, compared to those exerted by Bt, and thus delay the emergence of 'RNAi-resistant' insects. The possibility that insects currently sensitive to RNAi will evolve resistance cannot be excluded, and natural variation of proteins that are involved in dsRNA uptake and systemic RNAi between different insects may contribute to this. Thus, the appearance of insects resistant to RNAi could happen, but we expect it to be on a longer timescale. Strategies that reduce or delay the evolution of resistance, such as those used for transgenic Bt crops (Tabashnik *et al.*, 2005) should be used. Despite these drawbacks, we think that elevating the dsRNA expression level in plants and selecting more effective and specific target genes are urgently needed for further development of RNAi technology for insect-proof plants.

7 Concluding remarks

In short, RNAi as a sequence-specific gene silencing tool shows great potential for crop protection. This technology can be particularly valuable against those insects that are insensitive to current Bt crops, such as sucking insects and leafhoppers. A better understanding of plant and insect RNAi molecular mechanisms and plant–insect interactions will greatly help us to further improve this new generation of pest control technology. We believe that with continuous efforts toward increasing gene silencing efficacy and specificity, RNAi will work well in a wide range of agricultural applications, including crop protection.

Acknowledgements

We thank Chunmin Shan for manuscript preparation and Dr. Xiangxiong Zhu for helpful discussions. We are grateful to Jason R. Kennerdell, Richard W. Carthew, Yoshinori Tomoyasu, Robin E. Denell, James A Baum and James Roberts for providing pictures. The research is supported by National Science Foundation of China (30870380), The National High-tech Research Program of China (2009AA10Z112) and The Ministry of Agriculture of China (2008ZX08009-001, 2008ZX08005-001).

References

Abel, P. P., Nelson, R. S., De, B., Hoffmann, N., Rogers, S. G., Fraley, R. T. and Beachy, R. N. (1986). Delay of disease development in transgenic plants that express the tobacco mosaic virus coat protein gene. *Science* **232**, 738–743.

Aljamali, M., Bowman, A. S., Dillwith, J. W., Tucker, J. S., Yates, G. W., Essenberg, R. C. and Sauer, J. R. (2002). Identity and synthesis of prostaglandins in the lone star tick, Amblyomma americanum (L.), as assessed by radio-immunoassay and gas chromatography/mass spectrometry. *Insect Biochem. Mol. Biol.* **32**, 331–341.

Aljamali, M. N., Bior, A. D., Sauer, J. R. and Essenberg, R. C. (2003). RNA interference in ticks: a study using histamine binding protein dsRNA in the female tick Amblyomma americanum. *Insect Mol. Biol.* **12**, 299–305.

Allen, E., Xie, Z., Gustafson, A. M. and Carrington, J. C. (2005). microRNA-directed phasing during trans-acting siRNA biogenesis in plants. *Cell* **121**, 207–221.

Amparyup, P., Wiriyaukaradecha, K., Charoensapsri, W. and Tassanakajon, A. (2009). A clip domain serine proteinase plays a role in antibacterial defense but is not required for prophenoloxidase activation in shrimp. *Dev. Comp. Immunol.* **34**, 168–176.

Angenent, G. C., Franken, J., Busscher, M., Weiss, D. and van Tunen, A. J. (1994). Co-suppression of the petunia homeotic gene fbp2 affects the identity of the generative meristem. *Plant J.* **5**, 33–44.

Arakane, Y., Li, B., Muthukrishnan, S., Beeman, R. W., Kramer, K. J. and Park, Y. (2008). Functional analysis of four neuropeptides, EH, ETH, CCAP and bursicon, and their receptors in adult ecdysis behavior of the red flour beetle, Tribolium castaneum. *Mech. Dev.* **125**, 984–995.

Araujo, R. N., Santos, A., Pinto, F. S., Gontijo, N. F., Lehane, M. J. and Pereira, M. H. (2006). RNA interference of the salivary gland nitrophorin 2 in the triatomine bug Rhodnius prolixus (Hemiptera: Reduviidae) by dsRNA ingestion or injection. *Insect Biochem. Mol. Biol.* **36**, 683–693.

Aravin, A. A., Naumova, N. M., Tulin, A. V., Vagin, V. V., Rozovsky, Y. M. and Gvozdev, V. A. (2001). Double-stranded RNA-mediated silencing of genomic tandem repeats and transposable elements in the *D. melanogaster* germline. *Curr. Biol.* **11**, 1017–1027.

Aravin, A. A., Hannon, G. J. and Brennecke, J. (2007). The Piwi-piRNA pathway provides an adaptive defense in the transposon arms race. *Science* **318**, 761–764.

Aung, K., Boldbaatar, D., Liao, M., Umemiya-Shirafuji, R., Nakao, S., Matsuoka, T., Tanaka, T. and Fujisaki, K. (2011). Identification and characterization of class B scavenger receptor CD36 from the hard tick, *Haemaphysalis longicornis*. *Parasitol. Res.* **108**, 273–285.

Bagla, P. (2010). Hardy cotton-munching pests are latest blow to GM crops. *Science* **327**, 1439.

Bass, B. L. (2000). Double-stranded RNA as a template for gene silencing. *Cell* **101**, 235–238.

Batista, P. J., Ruby, J. G., Claycomb, J. M., Chiang, R., Fahlgren, N., Kasschau, K. D., Chaves, D. A., Gu, W., Vasale, J. J., Duan, S., Conte, D. Jr. Luo, S., *et al.* (2008). PRG-1 and 21U-RNAs interact to form the piRNA complex required for fertility in *C. elegans*. *Mol. Cell* **31**, 67–78.

Baum, J. A., Bogaert, T., Clinton, W., Heck, G. R., Feldmann, P., Ilagan, O., Johnson, S., Plaetinck, G., Munyikwa, T., Pleau, M., Vaughn, T. and Roberts, J. (2007). Control of coleopteran insect pests through RNA interference. *Nat. Biotechnol.* **25**, 1322–1326.

Bejarano, F., Smibert, P. and Lai, E. C. (2009). miR-9a prevents apoptosis during wing development by repressing Drosophila LIM-only. *Dev. Biol.* **338**, 63–73.

Belles, X. (2010). Beyond Drosophila: RNAi in vivo and functional genomics in insects. *Annu. Rev. Entomol.* **55**, 111–128.

Berge, J. B., Feyereisen, R. and Amichot, M. (1998). Cytochrome P450 monooxygenases and insecticide resistance in insects. *Philos. Trans. R. Soc. Lond. B Biol. Sci.* **353**, 1701–1705.

Biryukova, I., Asmar, J., Abdesselem, H. and Heitzler, P. (2009). Drosophila mir-9a regulates wing development via fine-tuning expression of the LIM only factor, dLMO. *Dev. Biol.* **327**, 487–496.

Boisson, B., Jacques, J. C., Choumet, V., Martin, E., Xu, J., Vernick, K. and Bourgouin, C. (2006). Gene silencing in mosquito salivary glands by RNAi. *FEBS Lett.* **580**, 1988–1992.

Bowman, A. S., Coons, L. B., Needham, G. R. and Sauer, J. R. (1997). Tick saliva: recent advances and implications for vector competence. *Med. Vet. Entomol.* **11**, 277–285.

Brennecke, J., Aravin, A. A., Stark, A., Dus, M., Kellis, M., Sachidanandam, R. and Hannon, G. J. (2007). Discrete small RNA-generating loci as master regulators of transposon activity in Drosophila. *Cell* **128**, 1089–1103.

Brosnan, C. A., Mitter, N., Christie, M., Smith, N. A., Waterhouse, P. M. and Carroll, B. J. (2007). Nuclear gene silencing directs reception of long-distance mRNA silencing in Arabidopsis. *Proc. Natl. Acad. Sci. USA* **104**, 14741–14746.

Brown, S. J., Mahaffey, J. P., Lorenzen, M. D., Denell, R. E. and Mahaffey, J. W. (1999). Using RNAi to investigate orthologous homeotic gene function during development of distantly related insects. *Evol. Dev.* **1**, 11–15.

Brusslan, J. A., Karlin-Neumann, G. A., Huang, L. and Tobin, E. M. (1993). An Arabidopsis mutant with a reduced level of cab140 RNA is a result of cosuppression. *Plant Cell* **5**, 667–677.

Bucher, G., Scholten, J. and Klingler, M. (2002). Parental RNAi in Tribolium (Coleoptera). *Curr. Biol.* **12**, R85–R86.

Cai, Y., Yu, X., Zhou, Q., Yu, C., Hu, H., Liu, J., Lin, H., Yang, J., Zhang, B., Cui, P., Hu, S. and Yu, J. (2010). Novel microRNAs in silkworm (Bombyx mori). *Funct. Integr. Genomics* **10**, 405–415.

Carthew, R. W. and Sontheimer, E. J. (2009). Origins and mechanisms of miRNAs and siRNAs. *Cell* **136**, 642–655.

Catalanotto, C., Azzalin, G., Macino, G. and Cogoni, C. (2000). Gene silencing in worms and fungi. *Nature* **404**, 245.

Cerutti, L., Mian, N. and Bateman, A. (2000). Domains in gene silencing and cell differentiation proteins: the novel PAZ domain and redefinition of the Piwi domain. *Trends Biochem. Sci.* **25**, 481–482.

Chan, S. W., Zilberman, D., Xie, Z., Johansen, L. K., Carrington, J. C. and Jacobsen, S. E. (2004). RNA silencing genes control de novo DNA methylation. *Science* **303**, 1336.

Chapman, E. J. and Carrington, J. C. (2007). Specialization and evolution of endogenous small RNA pathways. *Nat. Rev. Genet.* **8**, 884–896.

Chen, X., Yu, X., Cai, Y., Zheng, H., Yu, D., Liu, G., Zhou, Q., Hu, S. and Hu, F. (2010). Next-generation small RNA sequencing for microRNAs profiling in the honey bee Apis mellifera. *Insect Mol. Biol.* doi:10.1111/j.1365-2583.2010.01039.x.

Cifuentes, D., Xue, H., Taylor, D. W., Patnode, H., Mishima, Y., Cheloufi, S., Ma, E., Mane, S., Hannon, G. J., Lawson, N. D., Wolfe, S. A. and Giraldez, A. J. (2010). A novel miRNA processing pathway independent of dicer requires argonaute2 catalytic activity. *Science* **328**, 1694–1698.

Clemens, J. C., Worby, C. A., Simonson-Leff, N., Muda, M., Maehama, T., Hemmings, B. A. and Dixon, J. E. (2000). Use of double-stranded RNA interference in Drosophila cell lines to dissect signal transduction pathways. *Proc. Natl. Acad. Sci. USA* **97**, 6499–6503.

Conine, C. C., Batista, P. J., Gu, W., Claycomb, J. M., Chaves, D. A., Shirayama, M. and Mello, C. C. (2010). Argonautes ALG-3 and ALG-4 are required for spermatogenesis-specific 26G-RNAs and thermotolerant sperm in *Caenorhabditis elegans*. *Proc. Natl. Acad. Sci. USA* **107**, 3588–3593.

Crete, P., Leuenberger, S., Iglesias, V. A., Suarez, V., Schob, H., Holtorf, H., van Eeden, S. and Meins, F. (2001). Graft transmission of induced and spontaneous post-transcriptional silencing of chitinase genes. *Plant J.* **28**, 493–501.

Czech, B. and Hannon, G. J. (2011). Small RNA sorting: matchmaking for Argonautes. *Nat. Rev. Genet.* **12**, 19–31.

Czech, B., Malone, C. D., Zhou, R., Stark, A., Schlingeheyde, C., Dus, M., Perrimon, N., Kellis, M., Wohlschlegel, J. A., Sachidanandam, R., Hannon, G. J. and Brennecke, J. (2008). An endogenous small interfering RNA pathway in Drosophila. *Nature* **453**, 798–802.

Czech, B., Zhou, R., Erlich, Y., Brennecke, J., Binari, R., Villalta, C., Gordon, A., Perrimon, N. and Hannon, G. J. (2009). Hierarchical rules for Argonaute loading in Drosophila. *Mol. Cell* **36**, 445–456.

Dai, H., Ma, L., Wang, J., Jiang, R., Wang, Z. and Fei, J. (2008). Knockdown of ecdysis-triggering hormone gene with a binary UAS/GAL4 RNA interference system leads to lethal ecdysis deficiency in silkworm. *Acta Biochim. Biophys. Sin.* **40**, 790–795.

Das, P. P., Bagijn, M. P., Goldstein, L. D., Woolford, J. R., Lehrbach, N. J., Sapetschnig, A., Buhecha, H. R., Gilchrist, M. J., Howe, K. L., Stark, R., Matthews, N. Berezikov, E., *et al.* (2008). Piwi and piRNAs act upstream of an endogenous siRNA pathway to suppress Tc3 transposon mobility in the *Caenorhabditis elegans* germline. *Mol. Cell* **31**, 79–90.

De Carvalho, F., Gheysen, G., Kushnir, S., Van Montagu, M., Inze, D. and Castresana, C. (1992). Suppression of beta-1,3-glucanase transgene expression in homozygous plants. *EMBO J.* **11**, 2595–2602.

de Maagd, R. A., Bravo, A. and Crickmore, N. (2001). How Bacillus thuringiensis has evolved specific toxins to colonize the insect world. *Trends Genet.* **17**, 193–199.

Desset, S., Meignin, C., Dastugue, B. and Vaury, C. (2003). COM, a heterochromatic locus governing the control of independent endogenous retroviruses from *Drosophila melanogaster*. *Genetics* **164**, 501–509.

Dietzl, G., Chen, D., Schnorrer, F., Su, K. C., Barinova, Y., Fellner, M., Gasser, B., Kinsey, K., Oppel, S., Scheiblauer, S., Couto, A. Marra, V., *et al.* (2007). A genome-wide transgenic RNAi library for conditional gene inactivation in Drosophila. *Nature* **448**, 151–156.

Ding, X., Gopalakrishnan, B., Johnson, L. B., White, F. F., Wang, X., Morgan, T. D., Kramer, K. J. and Muthukrishnan, S. (1998). Insect resistance of transgenic tobacco expressing an insect chitinase gene. *Transgenic Res.* **7**, 77–84.

Du, L., Ge, F., Zhu, S. and Parajulee, M. N. (2004). Effect of cotton cultivar on development and reproduction of Aphis gossypii (Homoptera: Aphididae) and its predator Propylaea japonica (Coleoptera: Coccinellidae). *J. Econ. Entomol.* **97**, 1278–1283.

Dufourmantel, N., Pelissier, B., Garcon, F., Peltier, G., Ferullo, J. M. and Tissot, G. (2004). Generation of fertile transplastomic soybean. *Plant Mol. Biol.* **55**, 479–489.

Dunoyer, P., Himber, C. and Voinnet, O. (2005). DICER-LIKE 4 is required for RNA interference and produces the 21-nucleotide small interfering RNA component of the plant cell-to-cell silencing signal. *Nat. Genet.* **37**, 1356–1360.

Dunoyer, P., Himber, C., Ruiz-Ferrer, V., Alioua, A. and Voinnet, O. (2007). Intra- and intercellular RNA interference in Arabidopsis thaliana requires components of the microRNA and heterochromatic silencing pathways. *Nat. Genet.* **39**, 848–856.

Dunoyer, P., Brosnan, C. A., Schott, G., Wang, Y., Jay, F., Alioua, A., Himber, C. and Voinnet, O. (2010). An endogenous, systemic RNAi pathway in plants. *EMBO J.* **29**, 1699–1712.

Eguchi, A., Meade, B. R., Chang, Y. C., Fredrickson, C. T., Willert, K., Puri, N. and Dowdy, S. F. (2009). Efficient siRNA delivery into primary cells by a peptide transduction domain-dsRNA binding domain fusion protein. *Nat. Biotechnol.* **27**, 567–571.

Fahlgren, N., Montgomery, T. A., Howell, M. D., Allen, E., Dvorak, S. K., Alexander, A. L. and Carrington, J. C. (2006). Regulation of AUXIN RESPONSE FACTOR3 by TAS3 ta-siRNA affects developmental timing and patterning in Arabidopsis. *Curr. Biol.* **16**, 939–944.

Feinberg, E. H. and Hunter, C. P. (2003). Transport of dsRNA into cells by the transmembrane protein SID-1. *Science* **301**, 1545–1547.

Feldhaar, H. and Gross, R. (2009). Insects as hosts for mutualistic bacteria. *Int. J. Med. Microbiol.* **299**, 1–8.

Felgner, P. L., Gadek, T. R., Holm, M., Roman, R., Chan, H. W., Wenz, M., Northrop, J. P., Ringold, G. M. and Danielsen, M. (1987). Lipofection: a highly efficient, lipid-mediated DNA-transfection procedure. *Proc. Natl. Acad. Sci. USA* **84**, 7413–7417.

Fire, A., Xu, S., Montgomery, M. K., Kostas, S. A., Driver, S. E. and Mello, C. C. (1998). Potent and specific genetic interference by double-stranded RNA in *Caenorhabditis elegans*. *Nature* **391**, 806–811.

Fitches, E., Woodhouse, S. D., Edwards, J. P. and Gatehouse, J. A. (2001). In vitro and in vivo binding of snowdrop (Galanthus nivalis agglutinin; GNA) and jackbean (Canavalia ensiformis; Con A) lectins within tomato moth (Lacanobia oleracea) larvae; mechanisms of insecticidal action. *J. Insect Physiol.* **47**, 777–787.

Förstemann, K., Horwich, M. D., Wee, L., Tomari, Y. and Zamore, P. D. (2007). Drosophila microRNAs are sorted into functionally distinct argonaute complexes after production by dicer-1. *Cell* **130**, 287–297.

Gasciolli, V., Mallory, A. C., Bartel, D. P. and Vaucheret, H. (2005). Partially redundant functions of Arabidopsis DICER-like enzymes and a role for DCL4 in producing trans-acting siRNAs. *Curr. Biol.* **15**, 1494–1500.

Ghildiyal, M. and Zamore, P. D. (2009). Small silencing RNAs: an expanding universe. *Nat. Rev. Genet.* **10**, 94–108.

Gordon, K. H. J. and Waterhouse, P. M. (2007). RNAi for insect-proof plants. *Nat. Biotechnol.* **25**, 1231–1232.

Grosshans, H. and Filipowicz, W. (2008). Molecular biology: the expanding world of small RNAs. *Nature* **451**, 414–416.

Gunawardane, L. S., Saito, K., Nishida, K. M., Miyoshi, K., Kawamura, Y., Nagami, T., Siomi, H. and Siomi, M. C. (2007). A slicer-mediated mechanism for repeat-associated siRNA 5' end formation in Drosophila. *Science* **315**, 1587–1590.

Guo, S. and Kemphues, K. J. (1995). par-1, a gene required for establishing polarity in *C. elegans* embryos, encodes a putative Ser/Thr kinase that is asymmetrically distributed. *Cell* **81**, 611–620.

Hamilton, A. J. and Baulcombe, D. C. (1999). A species of small antisense RNA in posttranscriptional gene silencing in plants. *Science* **286**, 950–952.

Hammond, S. M., Bernstein, E., Beach, D. and Hannon, G. J. (2000). An RNA-directed nuclease mediates post-transcriptional gene silencing in Drosophila cells. *Nature* **404**, 293–296.

Hartig, J. V., Tomari, Y. and Forstemann, K. (2007). piRNAs—the ancient hunters of genome invaders. *Genes Dev.* **21**, 1707–1713.

Harvey, J. J. W., Lewsey, M. G., Patel, K., Westwood, J., Heimstädt, S., Carr, J. P. and Baulcombe, D. C. (2011). An antiviral defense role of AGO2 in plants. *PLoS One* **6**, e14639.

Havecker, E. R., Wallbridge, L. M., Hardcastle, T. J., Bush, M. S., Kelly, K. A., Dunn, R. M., Schwach, F., Doonan, J. H. and Baulcombe, D. C. (2010). The arabidopsis RNA-directed DNA methylation argonautes functionally diverge based on their expression and interaction with target loci. *Plant Cell* **22**, 321–334.

Henderson, I. R., Zhang, X., Lu, C., Johnson, L., Meyers, B. C., Green, P. J. and Jacobsen, S. E. (2006). Dissecting Arabidopsis thaliana DICER function in small RNA processing, gene silencing and DNA methylation patterning. *Nat. Genet.* **38**, 721–725.

Honeybee Genome Sequencing Consortium (2006). Insights into social insects from the genome of the honeybee *Apis mellifera*. *Nature* **444**, 512.

Hou, B.-K., Zhou, Y.-H., Wan, L.-H., Zhang, Z.-L., Shen, G.-F., Chen, Z.-H. and Hu, Z.-M. (2003). Chloroplast transformation in oilseed rape. *Transgenic Res.* **12**, 111–114.

Houwing, S., Kamminga, L. M., Berezikov, E., Cronembold, D., Girard, A., van den Elst, H., Filippov, D. V., Blaser, H., Raz, E., Moens, C. B., Plasterk, R. H. Hannon, G. J., *et al.* (2007). A role for Piwi and piRNAs in germ cell maintenance and transposon silencing in Zebrafish. *Cell* **129**, 69–82.

Huang, G., Allen, R., Davis, E., Baum, T. and Hussey, R. (2006). Engineering broad root-knot resistance in transgenic plants by RNAi silencing of a conserved and essential root-knot nematode parasitism gene. *Proc. Natl. Acad. Sci. USA* **103**, 14302–14306.

Huang, J., Zhang, Y., Li, M., Wang, S., Liu, W., Couble, P., Zhao, G. and Huang, Y. (2007). RNA interference-mediated silencing of the bursicon gene induces defects in wing expansion of silkworm. *FEBS Lett.* **581**, 697–701.

Hughes, C. L. and Kaufman, T. C. (2000). RNAi analysis of Deformed, proboscipedia and Sex combs reduced in the milkweed bug Oncopeltus fasciatus: novel roles for Hox genes in the hemipteran head. *Development* **127**, 3683–3694.

Hussain, M. and Asgari, S. (2009). Functional analysis of a cellular microRNA in insect host-ascovirus interaction. *J. Virol.* **84**, 612–620.

Hutvagner, G. and Simard, M. J. (2008). Argonaute proteins: key players in RNA silencing. *Nat. Rev. Mol. Cell Biol.* **9**, 22–32.

Huvenne, H. and Smagghe, G. (2010). Mechanisms of dsRNA uptake in insects and potential of RNAi for pest control: a review. *J. Insect Physiol.* **56**, 227–235.

Iovino, N., Pane, A. and Gaul, U. (2009). miR-184 has multiple roles in Drosophila female germline development. *Dev. Cell* **17**, 123–133.

Isoe, J., Rascon, A. A. Jr., Kunz, S. and Miesfeld, R. L. (2009). Molecular genetic analysis of midgut serine proteases in Aedes aegypti mosquitoes. *Insect Biochem. Mol. Biol.* **39**, 903–912.

Jagadeeswaran, G., Zheng, Y., Sumathipala, N., Jiang, H., Arrese, E. L., Soulages, J. L., Zhang, W. and Sunkar, R. (2010). Deep sequencing of small RNA libraries reveals dynamic regulation of conserved and novel microRNAs and microRNA-stars during silkworm development. *BMC Genomics* **11**, 52.

Jaubert-Possamai, S., Le Trionnaire, G., Bonhomme, J., Christophides, G., Rispe, C. and Tagu, D. (2007). Gene knockdown by RNAi in the pea aphid Acyrthosiphon pisum. *BMC Biotechnol.* **7**, 63.

Jose, A. M., Smith, J. J. and Hunter, C. P. (2009). Export of RNA silencing from *C. elegans* tissues does not require the RNA channel SID-1. *Proc. Natl. Acad. Sci. USA* **106**, 2283–2288.

Kanginakudru, S., Royer, C., Edupalli, S. V., Jalabert, A., Mauchamp, B., Chandrashekaraiah, , Prasad, S. V., Chavancy, G., Couble, P. and Nagaraju, J. (2007). Targeting ie-1 gene by RNAi induces baculoviral resistance in lepidopteran cell lines and in transgenic silkworms. *Insect Mol. Biol.* **16**, 635–644.

Katiyar-Agarwal, S., Morgan, R., Dahlbeck, D., Borsani, O., Villegas, A., Zhu, J.-K., Staskawicz, B. J. and Jin, H. (2006). A pathogen-inducible endogenous siRNA in plant immunity. *Proc. Natl. Acad. Sci. USA* **103**, 18002–18007.

Katiyar-Agarwal, S., Gao, S., Vivian-Smith, A. and Jin, H. (2007). A novel class of bacteria-induced small RNAs in Arabidopsis. *Genes Dev.* **21**, 3123–3134.
Kennerdell, J. R. and Carthew, R. W. (1998). Use of dsRNA-mediated genetic interference to demonstrate that frizzled and frizzled 2 act in the wingless pathway. *Cell* **95**, 1017–1026.
Kennerdell, J. R. and Carthew, R. W. (2000). Heritable gene silencing in Drosophila using double-stranded RNA. *Nat. Biotechnol.* **18**, 896–898.
Kocks, C., Cho, J. H., Nehme, N., Ulvila, J., Pearson, A. M., Meister, M., Strom, C., Conto, S. L., Hetru, C., Stuart, L. M., Stehle, T. Hoffmann, J. A., *et al.* (2005). Eater, a transmembrane protein mediating phagocytosis of bacterial pathogens in Drosophila. *Cell* **123**, 335–346.
Krause, P. J., McKay, K., Thompson, C. A., Sikand, V. K., Lentz, R., Lepore, T., Closter, L., Christianson, D., Telford, S. R., Persing, D., Radolf, J. D. and Spielman, A. (2002). Disease-specific diagnosis of coinfecting tickborne zoonoses: babesiosis, human granulocytic ehrlichiosis, and Lyme disease. *Clin. Infect. Dis.* **34**, 1184–1191.
Kumar, S., Dhingra, A. and Daniell, H. (2004). Stable transformation of the cotton plastid genome and maternal inheritance of transgenes. *Plant Mol. Biol.* **56**, 203–216.
Lee, Y. S., Nakahara, K., Pham, J. W., Kim, K., He, Z., Sontheimer, E. J. and Carthew, R. W. (2004). Distinct roles for Drosophila Dicer-1 and Dicer-2 in the siRNA/miRNA silencing pathways. *Cell* **117**, 69–81.
Legeai, F., Rizk, G., Walsh, T., Edwards, O., Gordon, K., Lavenier, D., Leterme, N., Mereau, A., Nicolas, J., Tagu, D. and Jaubert-Possamai, S. (2010). Bioinformatic prediction, deep sequencing of microRNAs and expression analysis during phenotypic plasticity in the pea aphid, Acyrthosiphon pisum. *BMC Genomics* **11**, 281.
Li, W., Zangerl, A. R., Schuler, M. A. and Berenbaum, M. R. (2004). Characterization and evolution of furanocoumarin-inducible cytochrome P450s in the parsnip webworm, *Depressaria pastinacella. Insect Mol. Biol.* **13**, 603–613.
Li, Y., Wang, F., Lee, J. A. and Gao, F. B. (2006). MicroRNA-9a ensures the precise specification of sensory organ precursors in Drosophila. *Genes Dev.* **20**, 2793–2805.
Li, S., Mead, E. A., Liang, S. and Tu, Z. (2009). Direct sequencing and expression analysis of a large number of miRNAs in Aedes aegypti and a multi-species survey of novel mosquito miRNAs. *BMC Genomics* **10**, 581.
Lin, H. (2007). piRNAs in the germ line. *Science* **316**, 397.
Lindbo, J. A. and Dougherty, W. G. (1992). Untranslatable transcripts of the tobacco etch virus coat protein gene sequence can interfere with tobacco etch virus replication in transgenic plants and protoplasts. *Virology* **189**, 725–733.
Lingel, A., Simon, B., Izaurralde, E. and Sattler, M. (2003). Structure and nucleic-acid binding of the Drosophila Argonaute 2 PAZ domain. *Nature* **426**, 465–469.
Lipardi, C. and Paterson, B. M. (2009). Identification of an RNA-dependent RNA polymerase in Drosophila involved in RNAi and transposon suppression. *Proc. Natl. Acad. Sci. USA* **106**, 15645–15650.
Lippman, Z. and Martienssen, R. (2004). The role of RNA interference in heterochromatic silencing. *Nature* **431**, 364–370.
Liu, Q., Rand, T. A., Kalidas, S., Du, F., Kim, H. E., Smith, D. P. and Wang, X. (2003). R2D2, a bridge between the initiation and effector steps of the Drosophila RNAi pathway. *Science* **301**, 1921–1925.
Liu, C. W., Lin, C. C., Chen, J. J. and Tseng, M. J. (2007). Stable chloroplast transformation in cabbage (Brassica oleracea L. var. capitata L.) by particle bombardment. *Plant Cell Rep.* **26**, 1733–1744.
Liu, C., Yamamoto, K., Cheng, T.-C., Kadono-Okuda, K., Narukawa, J., Liu, S.-P., Han, Y., Futahashi, R., Kidokoro, K., Noda, H., Kobayashi, I. Tamura, T., *et al.*

(2010a). Repression of tyrosine hydroxylase is responsible for the sex-linked chocolate mutation of the silkworm, Bombyx mori. *Proc. Natl. Acad. Sci. USA* **107**, 12980–12985.

Liu, S., Ding, Z., Zhang, C., Yang, B. and Liu, Z. (2010b). Gene knockdown by intro-thoracic injection of double-stranded RNA in the brown planthopper, *Nilaparvata lugens*. *Insect Biochem. Mol. Biol.* **40**, 666–671.

Lu, Y., Wu, K., Jiang, Y., Xia, B., Li, P., Feng, H., Wyckhuys, K. A. and Guo, Y. (2010). Mirid bug outbreaks in multiple crops correlated with wide-scale adoption of Bt cotton in China. *Science* **328**, 1151–1154.

Luo, P., Wang, Y. H., Wang, G. D., Essenberg, M. and Chen, X. Y. (2001). Molecular cloning and functional identification of (+)-delta-cadinene-8-hydroxylase, a cytochrome P450 mono-oxygenase (CYP706B1) of cotton sesquiterpene biosynthesis. *Plant J.* **28**, 95–104.

Maleszka, J., Forêt, S., Saint, R. and Maleszka, R. (2007). RNAi-induced phenotypes suggest a novel role for a chemosensory protein CSP5 in the development of embryonic integument in the honeybee (Apis mellifera). *Dev. Genes Evol.* **217**, 189–196.

Mallory, A. C., Hinze, A., Tucker, M. R., Bouché, N., Gasciolli, V., Elmayan, T., Lauressergues, D., Jauvion, V., Vaucheret, H. and Laux, T. (2009). Redundant and specific roles of the ARGONAUTE proteins AGO1 and ZLL in development and small RNA-directed gene silencing. *PLoS Genet.* **5**, e1000646.

Malone, R. W., Felgner, P. L. and Verma, I. M. (1989). Cationic liposome-mediated RNA transfection. *Proc. Natl. Acad. Sci. USA* **86**, 6077–6081.

Mao, Y. B., Cai, W. J., Wang, J. W., Hong, G. J., Tao, X. Y., Wang, L. J., Huang, Y. P. and Chen, X. Y. (2007). Silencing a cotton bollworm P450 monooxygenase gene by plant-mediated RNAi impairs larval tolerance of gossypol. *Nat. Biotechnol.* **25**, 1307–1313.

Mao, Y. B., Tao, X. Y., Xue, X. Y., Wang, L. J. and Chen, X. Y. (2011). Cotton plants expressing CYP6AE14 double-stranded RNA show enhanced resistance to bollworms. *Transgenic Res.* **20**, 665–673.

Martinez, J., Patkaniowska, A., Urlaub, H., Luhrmann, R. and Tuschl, T. (2002). Single-stranded antisense siRNAs guide target RNA cleavage in RNAi. *Cell* **110**, 563–574.

Marvier, M., McCreedy, C., Regetz, J. and Kareiva, P. (2007). A meta-analysis of effects of Bt cotton and maize on nontarget invertebrates. *Science* **316**, 1475–1477.

Matzke, M. A. and Matzke, A. J. (1991). Differential inactivation and methylation of a transgene in plants by two suppressor loci containing homologous sequences. *Plant Mol. Biol.* **16**, 821–830.

Mazumdar-Leighton, S. and Broadway, R. M. (2001). Identification of six chymotrypsin cDNAs from larval midguts of Helicoverpa zea and Agrotis ipsilon feeding on the soybean (Kunitz) trypsin inhibitor. *Insect Biochem. Mol. Biol.* **31**, 633–644.

McBride, K. E., Svab, Z., Schaaf, D. J., Hogan, P. S., Stalker, D. M. and Maliga, P. (1995). Amplification of a chimeric Bacillus gene in chloroplasts leads to an extraordinary level of an insecticidal protein in tobacco. *Biotechnology (N. Y)* **13**, 362–365.

Mi, S., Cai, T., Hu, Y., Chen, Y., Hodges, E., Ni, F., Wu, L., Li, S., Zhou, H., Long, C., Chen, S. Hannon, G. J., *et al.* (2008). Sorting of small RNAs into Arabidopsis argonaute complexes is directed by the 5' terminal nucleotide. *Cell* **133**, 116–127.

Miller, S., Brown, S. and Tomoyasu, Y. (2008). Larval RNAi in *Drosophila*? *Dev. Genes Evol.* **218**, 505–510.

Moazed, D. (2009). Small RNAs in transcriptional gene silencing and genome defence. *Nature* **457**, 413–420.

Napoli, C., Lemieux, C. and Jorgensen, R. (1990). Introduction of a chimeric chalcone synthase gene into petunia results in reversible co-suppression of homologous genes in trans. *Plant Cell* **2**, 279–289.

Narasimhan, S., Montgomery, R. R., DePonte, K., Tschudi, C., Marcantonio, N., Anderson, J. F., Sauer, J. R., Cappello, M., Kantor, F. S. and Fikrig, E. (2004). Disruption of Ixodes scapularis anticoagulation by using RNA interference. *Proc. Natl. Acad. Sci. USA* **101**, 1141–1146.

Newmark, P. A., Reddien, P. W., Cebria, F. and Sanchez Alvarado, A. (2003). Ingestion of bacterially expressed double-stranded RNA inhibits gene expression in planarians. *Proc. Natl. Acad. Sci. USA* **100**(Suppl. 1), 11861–11865.

O'Donnell, K. A. and Boeke, J. D. (2007). Mighty Piwis defend the germline against genome intruders. *Cell* **129**, 37–44.

Ohizumi, Y., Gaidamashvili, M., Ohwada, S., Matsuda, K., Kominami, J., Nakamura-Tsuruta, S., Hirabayashi, J., Naganuma, T., Ogawa, T. and Muramoto, K. (2009). Mannose-binding lectin from yam (Dioscorea batatas) tubers with insecticidal properties against Helicoverpa armigera (Lepidoptera: Noctuidae). *J. Agric. Food Chem.* **57**, 2896–2902.

Okamura, K., Ishizuka, A., Siomi, H. and Siomi, M. C. (2004). Distinct roles for Argonaute proteins in small RNA-directed RNA cleavage pathways. *Genes Dev.* **18**, 1655–1666.

Okamura, K., Balla, S., Martin, R., Liu, N. and Lai, E. C. (2008a). Two distinct mechanisms generate endogenous siRNAs from bidirectional transcription in Drosophila melanogaster. *Nat. Struct. Mol. Biol.* **15**, 581–590.

Okamura, K., Chung, W. J., Ruby, J. G., Guo, H., Bartel, D. P. and Lai, E. C. (2008b). The Drosophila hairpin RNA pathway generates endogenous short interfering RNAs. *Nature* **453**, 803–806.

Okamura, K., Liu, N. and Lai, E. C. (2009). Distinct mechanisms for microRNA strand selection by Drosophila Argonautes. *Mol. Cell* **36**, 431–444.

Okumura, S., Sawada, M., Park, Y. W., Hayashi, T., Shimamura, M., Takase, H. and Tomizawa, K. (2006). Transformation of poplar (Populus alba) plastids and expression of foreign proteins in tree chloroplasts. *Transgenic Res.* **15**, 637–646.

Onodera, Y., Haag, J. R., Ream, T., Nunes, P. C., Pontes, O. and Pikaard, C. S. (2005). Plant nuclear RNA polymerase IV mediates siRNA and DNA methylation-dependent heterochromatin formation. *Cell* **120**, 613–622.

Orii, H., Mochii, M. and Watanabe, K. (2003). A simple "soaking method" for RNA interference in the planarian Dugesia japonica. *Dev. Genes Evol.* **213**, 138–141.

Pal-Bhadra, M., Leibovitch, B. A., Gandhi, S. G., Rao, M., Bhadra, U., Birchler, J. A. and Elgin, S. C. (2004). Heterochromatic silencing and HP1 localization in Drosophila are dependent on the RNAi machinery. *Science* **303**, 669–672.

Pan, M. H., Wang, X. Y., Chai, C. L., Zhang, C. D., Lu, C. and Xiang, Z. H. (2009). Identification and function of Abdominal-A in the silkworm, *Bombyx mori*. *Insect Mol. Biol.* **18**, 155–160.

Parrish, S., Fleenor, J., Xu, S., Mello, C. and Fire, A. (2000). Functional anatomy of a dsRNA trigger: differential requirement for the two trigger strands in RNA interference. *Mol. Cell* **6**, 1077–1087.

Price, D. R. and Gatehouse, J. A. (2008). RNAi-mediated crop protection against insects. *Trends Biotechnol.* **26**, 393–400.

Prud'homme, N., Gans, M., Masson, M., Terzian, C. and Bucheton, A. (1995). Flamenco, a gene controlling the gypsy retrovirus of *Drosophila melanogaster*. *Genetics* **139**, 697–711.

Purkayastha, A. and Dasgupta, I. (2009). Virus-induced gene silencing: a versatile tool for discovery of gene functions in plants. *Plant Physiol. Biochem.* **47**, 967–976.

Qi, Y., He, X., Wang, X. J., Kohany, O., Jurka, J. and Hannon, G. J. (2006). Distinct catalytic and non-catalytic roles of ARGONAUTE4 in RNA-directed DNA methylation. *Nature* **443**, 1008–1012.

Rämet, M., Pearson, A., Manfruelli, P., Li, X., Koziel, H., Göbel, V., Chung, E., Krieger, M. and Ezekowitz, R. A. B. (2001). Drosophila scavenger receptor CI is a pattern recognition receptor for bacteria. *Immunity* **15**, 1027–1038.

Reinhart, B. J., Slack, F. J., Basson, M., Pasquinelli, A. E., Bettinger, J. C., Rougvie, A. E., Horvitz, H. R. and Ruvkun, G. (2000). The 21-nucleotide let-7 RNA regulates developmental timing in *Caenorhabditis elegans*. *Nature* **403**, 901–906.

Ruf, S., Hermann, M., Berger, I. J., Carrer, H. and Bock, R. (2001). Stable genetic transformation of tomato plastids and expression of a foreign protein in fruit. *Nat. Biotechnol.* **19**, 870–875.

Saito, K., Nishida, K. M., Mori, T., Kawamura, Y., Miyoshi, K., Nagami, T., Siomi, H. and Siomi, M. C. (2006). Specific association of Piwi with rasiRNAs derived from retrotransposon and heterochromatic regions in the Drosophila genome. *Genes Dev.* **20**, 2214–2222.

Saleh, M. C., van Rij, R. P., Hekele, A., Gillis, A., Foley, E., O'Farrell, P. H. and Andino, R. (2006). The endocytic pathway mediates cell entry of dsRNA to induce RNAi silencing. *Nat. Cell Biol.* **8**, 793–802.

Sandmann, T. and Cohen, S. M. (2007). Identification of novel *Drosophila melanogaster* microRNAs. *PLoS One* **2**, e1265.

Sanford, J. C. and Johnston, S. A. (1985). The concept of parasite-derived resistance—deriving resistance genes from the parasite's own genome. *J. Theor. Biol.* **113**, 395–405.

Scott, J. G. and Wen, Z. (2001). Cytochromes P450 of insects: the tip of the iceberg. *Pest Manag. Sci.* **57**, 958–967.

Scott, J. G., Liu, N. and Wen, Z. (1998). Insect cytochromes P450: diversity, insecticide resistance and tolerance to plant toxins. *Comp. Biochem. Physiol. C Pharmacol. Toxicol. Endocrinol.* **121**, 147–155.

Seinen, E., Burgerhof, J. G. M., Jansen, R. C. and Sibon, O. C. M. (2010). RNAi experiments in *D. melanogaster*: solutions to the overlooked problem of off-targets shared by independent dsRNAs. *PLoS One* **5**, e13119.

Shih, J. D., Fitzgerald, M. C., Sutherlin, M. and Hunter, C. P. (2009). The SID-1 double-stranded RNA transporter is not selective for dsRNA length. *RNA* **15**, 384–390.

Sidorov, V. A., Kasten, D., Pang, S. Z., Hajdukiewicz, P. T., Staub, J. M. and Nehra, N. S. (1999). Technical advance: stable chloroplast transformation in potato: use of green fluorescent protein as a plastid marker. *Plant J.* **19**, 209–216.

Sijen, T., Fleenor, J., Simmer, F., Thijssen, K. L., Parrish, S., Timmons, L., Plasterk, R. H. A. and Fire, A. (2001). On the role of RNA amplification in dsRNA-triggered gene silencing. *Cell* **107**, 465–476.

Sijen, T., Steiner, F. A., Thijssen, K. L. and Plasterk, R. H. A. (2007). Secondary siRNAs result from unprimed RNA synthesis and form a distinct class. *Science* **315**, 244–247.

Siomi, H. and Siomi, M. C. (2009). On the road to reading the RNA-interference code. *Nature* **457**, 396–404.

Smith, L. M., Pontes, O., Searle, I., Yelina, N., Yousafzai, F. K., Herr, A. J., Pikaard, C. S. and Baulcombe, D. C. (2007). An SNF2 protein associated with nuclear RNA silencing and the spread of a silencing signal between cells in Arabidopsis. *Plant Cell* **19**, 1507–1521.

Soares, C. A., Lima, C. M., Dolan, M. C., Piesman, J., Beard, C. B. and Zeidner, N. S. (2005). Capillary feeding of specific dsRNA induces silencing of the isac gene in nymphal Ixodes scapularis ticks. *Insect Mol. Biol.* **14**, 443–452.

Song, J. J. and Joshua-Tor, L. (2006). Argonaute and RNA—getting into the groove. *Curr. Opin. Struct. Biol.* **16**, 5–11.

Song, J. J., Liu, J., Tolia, N. H., Schneiderman, J., Smith, S. K., Martienssen, R. A., Hannon, G. J. and Joshua-Tor, L. (2003). The crystal structure of the Argonaute2 PAZ domain reveals an RNA binding motif in RNAi effector complexes. *Nat. Struct. Biol.* **10**, 1026–1032.

Song, J. J., Smith, S. K., Hannon, G. J. and Joshua-Tor, L. (2004). Crystal structure of Argonaute and its implications for RISC slicer activity. *Science* **305**, 1434–1437.

Stipanovic, R. D., Lopez, J. D. Jr., Dowd, M. K., Puckhaber, L. S. and Duke, S. E. (2006). Effect of racemic and (+)- and (−)-gossypol on the survival and development of Helicoverpa zea larvae. *J. Chem. Ecol.* **32**, 959–968.

Tabara, H., Grishok, A. and Mello, C. C. (1998). RNAi in *C. elegans*: soaking in the genome sequence. *Science* **282**, 430–431.

Tabara, H., Sarkissian, M., Kelly, W. G., Fleenor, J., Grishok, A., Timmons, L., Fire, A. and Mello, C. C. (1999). The rde-1 gene, RNA interference, and transposon silencing in *C. elegans*. *Cell* **99**, 123–132.

Tabashnik, B. E., Carrière, Y., Dennehy, T. J., Morin, S., Sisterson, M. S., Roush, R. T., Shelton, A. M. and Zhao, J. Z. (2003). Insect resistance to transgenic Bt crops: lessons from the laboratory and field. *J. Econ. Entomol.* **96**, 1031–1038.

Tabashnik, B. E., Dennehy, T. J. and Carrière, Y. (2005). Delayed resistance to transgenic cotton in pink bollworm. *Proc. Natl. Acad. Sci. USA* **102**, 15389–15393.

Tabashnik, B. E., Gassmann, A. J., Crowder, D. W. and Carriere, Y. (2008). Insect resistance to Bt crops: evidence versus theory. *Nat. Biotechnol.* **26**, 199–202.

Tan, A. and Palli, S. R. (2008). Identification and characterization of nuclear receptors from the red flour beetle, *Tribolium castaneum*. *Insect Biochem. Mol. Biol.* **38**, 430–439.

Tan, X. P., Liang, W. Q., Liu, C. J., Luo, P., Heinstein, P. and Chen, X. Y. (2000). Expression pattern of (+)-delta-cadinene synthase genes and biosynthesis of sesquiterpene aldehydes in plants of Gossypium arboreum L. *Planta* **210**, 644–651.

Terenius, O., Papanicolaou, A., Garbutt, J. S., Eleftherianos, I., Huvenne, H., Kanginakudru, S., Albrechtsen, M., An, C., Aymeric, J.-L., Barthel, A., Bebas, P. Bitra, K., *et al.* (2011). RNA interference in Lepidoptera: an overview of successful and unsuccessful studies and implications for experimental design. *J. Insect Physiol.* **57**, 231–245.

Tian, H., Peng, H., Yao, Q., Chen, H., Xie, Q., Tang, B. and Zhang, W. (2009). Developmental control of a lepidopteran pest Spodoptera exigua by ingestion of bacteria expressing dsRNA of a non-midgut gene. *PLoS One* **4**, e6225.

Tijsterman, M., May, R. C., Simmer, F., Okihara, K. L. and Plasterk, R. H. A. (2004). Genes required for systemic RNA interference in Caenorhabditis elegans. *Curr. Biol.* **14**, 111–116.

Timmons, L. and Fire, A. (1998). Specific interference by ingested dsRNA. *Nature* **395**, 854.

Timmons, L., Court, D. L. and Fire, A. (2001). Ingestion of bacterially expressed dsRNAs can produce specific and potent genetic interference in *Caenorhabditis elegans*. *Gene* **263**, 103–112.

Tomoyasu, Y. and Denell, R. E. (2004). Larval RNAi in Tribolium (Coleoptera) for analyzing adult development. *Dev. Genes Evol.* **214**, 575–578.

Tomoyasu, Y., Miller, S. C., Tomita, S., Schoppmeier, M., Grossmann, D. and Bucher, G. (2008). Exploring systemic RNA interference in insects: a genome-wide survey for RNAi genes in Tribolium. *Genome Biol.* **9**, R10.

Tribolium Genome Sequencing Consortium (2008). The genome of the model beetle and pest Tribolium castaneum. *Nature* **452**, 949–955.

Turner, C. T., Davy, M. W., MacDiarmid, R. M., Plummer, K. M., Birch, N. P. and Newcomb, R. D. (2006). RNA interference in the light brown apple moth, Epiphyas postvittana (Walker) induced by double-stranded RNA feeding. *Insect Mol. Biol.* **15**, 383–391.

Ulvila, J., Parikka, M., Kleino, A., Sormunen, R., Ezekowitz, R. A., Kocks, C. and Ramet, M. (2006). Double-stranded RNA is internalized by scavenger receptor-mediated endocytosis in Drosophila S2 cells. *J. Biol. Chem.* **281**, 14370–14375.

Ulvila, J., Hultmark, D. and Ramet, M. (2010). RNA silencing in the antiviral innate immune defence—role of DEAD-box RNA helicases. *Scand. J. Immunol.* **71**, 146–158.

Vagin, V. V., Sigova, A., Li, C., Seitz, H., Gvozdev, V. and Zamore, P. D. (2006). A distinct small RNA pathway silences selfish genetic elements in the germline. *Science* **313**, 320–324.

Valadi, H., Ekstrom, K., Bossios, A., Sjostrand, M., Lee, J. J. and Lotvall, J. O. (2007). Exosome-mediated transfer of mRNAs and microRNAs is a novel mechanism of genetic exchange between cells. *Nat. Cell Biol.* **9**, 654–659.

Vallier, A., Vincent-Monegat, C., Laurencon, A. and Heddi, A. (2009). RNAi in the cereal weevil Sitophilus spp: systemic gene knockdown in the bacteriome tissue. *BMC Biotechnol.* **9**, 44.

Vargas-Vila, M. A., Hannibal, R. L., Parchem, R. J., Liu, P. Z. and Patel, N. H. (2010). A prominent requirement for single-minded and the ventral midline in patterning the dorsoventral axis of the crustacean Parhyale hawaiensis. *Development* **137**, 3469–3476.

Vazquez, F. (2006). Arabidopsis endogenous small RNAs: highways and byways. *Trends Plant Sci.* **11**, 460–468.

Vazquez, F., Vaucheret, H., Rajagopalan, R., Lepers, C., Gasciolli, V., Mallory, A. C., Hilbert, J. L., Bartel, D. P. and Crete, P. (2004). Endogenous trans-acting siRNAs regulate the accumulation of Arabidopsis mRNAs. *Mol. Cell* **16**, 69–79.

Verdel, A., Jia, S., Gerber, S., Sugiyama, T., Gygi, S., Grewal, S. I. and Moazed, D. (2004). RNAi-mediated targeting of heterochromatin by the RITS complex. *Science* **303**, 672–676.

Voinnet, O., Vain, P., Angell, S. and Baulcombe, D. C. (1998). Systemic spread of sequence-specific transgene RNA degradation in plants is initiated by localized introduction of ectopic promoterless DNA. *Cell* **95**, 177–187.

Waterhouse, P. M., Graham, M. W. and Wang, M.-B. (1998). Virus resistance and gene silencing in plants can be induced by simultaneous expression of sense and antisense RNA. *Proc. Natl. Acad. Sci. USA* **95**, 13959–13964.

Whyard, S., Singh, A. D. and Wong, S. (2009). Ingested double-stranded RNAs can act as species-specific insecticides. *Insect Biochem. Mol. Biol.* **39**, 824–832.

Wilson, A. C. C., Ashton, P. D., Calevro, F., Charles, H., Colella, S., Febvay, G., Jander, G., Kushlan, P. F., Macdonald, S. J., Schwartz, J. F., Thomas, G. H. and Douglas, A. E. (2010). Genomic insight into the amino acid relations of the pea aphid, Acyrthosiphon pisum, with its symbiotic bacterium Buchnera aphidicola. *Insect Mol. Biol.* **19**, 249–258.

Winston, W. M., Molodowitch, C. and Hunter, C. P. (2002). Systemic RNAi in *C. elegans* requires the putative transmembrane protein SID-1. *Science* **295**, 2456–2459.

Winston, W. M., Sutherlin, M., Wright, A. J., Feinberg, E. H. and Hunter, C. P. (2007). Caenorhabditis elegans SID-2 is required for environmental RNA interference. *Proc. Natl. Acad. Sci. USA* **104**, 10565–10570.

Wittstock, U., Agerbirk, N., Stauber, E. J., Olsen, C. E., Hippler, M., Mitchell-Olds, T., Gershenzon, J. and Vogel, H. (2004). Successful herbivore attack due to metabolic diversion of a plant chemical defense. *Proc. Natl. Acad. Sci. USA* **101**, 4859–4864.

Wolfrum, C., Shi, S., Jayaprakash, K. N., Jayaraman, M., Wang, G., Pandey, R. K., Rajeev, K. G., Nakayama, T., Charrise, K., Ndungo, E. M., Zimmermann, T. Koteliansky, V., *et al.* (2007). Mechanisms and optimization of *in vivo* delivery of lipophilic siRNAs. *Nat. Biotechnol.* **25**, 1149–1157.

Xiang, H., Li, M., Yang, F., Guo, Q., Zhan, S., Lin, H., Miao, X. and Huang, Y. (2008). Fine mapping of Ekp-1, a locus associated with silkworm (Bombyx mori) proleg development. *Heredity* **100**, 533–540.

Xiang, H., Li, M. W., Guo, J. H., Jiang, J. H. and Huang, Y. P. (2011). Influence of RNAi knockdown for E-complex genes on the silkworm proleg development. *Arch. Insect Biochem. Physiol.* **76**, 1–11.

Xie, Z., Johansen, L. K., Gustafson, A. M., Kasschau, K. D., Lellis, A. D., Zilberman, D., Jacobsen, S. E. and Carrington, J. C. (2004). Genetic and functional diversification of small RNA pathways in plants. *PLoS Biol.* **2**, 0642–0652.

Xu, W. and Han, Z. (2008). Cloning and phylogenetic analysis of sid-1-like genes from aphids. *J. Insect Sci.* **8**, 1–6.

Yigit, E., Batista, P. J., Bei, Y., Pang, K. M., Chen, C. C., Tolia, N. H., Joshua-Tor, L., Mitani, S., Simard, M. J. and Mello, C. C. (2006). Analysis of the *C. elegans* Argonaute family reveals that distinct Argonautes act sequentially during RNAi. *Cell* **127**, 747–757.

Yin, H. and Lin, H. (2007). An epigenetic activation role of Piwi and a Piwi-associated piRNA in *Drosophila melanogaster*. *Nature* **450**, 304–308.

Yu, X., Zhou, Q., Cai, Y., Luo, Q., Lin, H., Hu, S. and Yu, J. (2009). A discovery of novel microRNAs in the silkworm (Bombyx mori) genome. *Genomics* **94**, 438–444.

Zamore, P. D., Tuschl, T., Sharp, P. A. and Bartel, D. P. (2000). RNAi: double-stranded RNA directs the ATP-dependent cleavage of mRNA at 21 to 23 nucleotide intervals. *Cell* **101**, 25–33.

Zheng, X., Zhu, J., Kapoor, A. and Zhu, J. K. (2007). Role of Arabidopsis AGO6 in siRNA accumulation, DNA methylation and transcriptional gene silencing. *EMBO J.* **26**, 1691–1701.

Zilberman, D., Cao, X. and Jacobsen, S. E. (2003). ARGONAUTE4 control of locus-specific siRNA accumulation and DNA and histone methylation. *Science* **299**, 716–719.

RNA Interference in Ticks: A Functional Genomics Tool for the Study of Physiology

Shahid Karim and Steven W. Adamson

Department of Biological Sciences, University of Southern Mississippi, Hattiesburg, Mississippi, USA

ADVANCES IN INSECT PHYSIOLOGY VOL. 42
ISBN 978-0-12-387680-5
DOI: 10.1016/B978-0-12-387680-5.00004-5

Abstract

Ticks are an example of an efficient ectoparasite that feeds on a variety of hosts including humans, domestic and wild animals. Tick salivary glands are critical to the biological success of ticks both during extended periods off the host and during the feeding period on the host. The salivary glands are also the sites of pathogen development and saliva is the route of transmission. The importance of multifunctional salivary glands to tick survival and vector competency makes the glands potential targets for intervention. RNA interference (RNAi) has the potential to revolutionize genetic manipulation and enhance the development of therapeutic and control agents in many arthropod-borne diseases and vectors. This reverse genetics tool has already been successfully used to study tick–host and tick–pathogen interactions. The translation of RNAi from an effective functional genomics tool into field application has been hindered by the challenge of delivering RNAi molecules to their target tissues by systemic administration. This review highlights the current status of RNAi in defining the physiological role of tick molecules in vector competence and vector–host interactions.

1 Ticks and tick-borne diseases

Ticks are found in almost every region of the world and are second only to mosquitoes in their public health and veterinary importance (Sonenshine, 1991). A total of 870 tick species have been identified around the world (Wheller and Hayashi, 1998). Ticks are divided into two main families, Ixodidae (hard ticks) and Argasidae (soft ticks). A total of 683 hard and 183 soft tick species have been identified feasting on the blood of various host animals. The life cycle of a typical hard tick consists of four developmental stages: the egg, the six-legged larva, eight-legged nymph and adult (♂ or ♀). Typically, larvae and nymphs (undifferentiated sex stages) molt after feeding on nutritious blood into the next developmental stage. Female ticks feed on blood, mate with a male and drop off to lay eggs, completing the life cycle. The ixodid ticks take only one big blood meal during their prolonged feeding at each developmental stage, while argasid ticks feed intermittently on the same host (Parola and Raoult, 2001; Sonenshine, 1991). Ixodidae are divided into metastriate and prostriate ticks. Metastriate ticks have short mouthparts as compared to prostriate ticks. The ixodid ticks are generally classified based on their feeding preferences into one-, two- and three-host ticks. A single-host tick (e.g. *Rhipicephalus microplus/annulatus*) remains on a single host individual for feeding, molting, mating and engorgement in the adult stage. For two-host ticks (e.g. *Hyalomma* species), larvae and nymphs blood-feed on the first host individual and adults mate and engorge on the second host individual, which may be of different species. In three-host ticks (e.g. *Amblyomma*, *Ixodes* species), each developmental stage (larva, nymph and adult) blood feeds on a different mammalian host to complete the life cycle (Sonenshine, 1991). During blood feeding, ticks (e.g. *Amblyomma maculatum*) may inoculate the host with a toxin produced from the salivary glands, leading to a neuromuscular condition commonly known as tick paralysis

(Espinoza-Gomez *et al.*, 2011; Paffenbarger, 1951). Although rarely severe, tick paralysis may lead to muscular and respiratory failure and death (Cupp, 1991).

Ticks have a diverse range of vertebrate hosts and can inject and acquire a plethora of disease agents. Ticks transmit the greatest variety of pathogens to humans and veterinary species of any arthropod vector (Balashov, 1972), including the agents causing anaplasmosis (*Anaplasma phagocytophilum*, *A. marginale*), cowdriosis (*Ehrlichia ruminantium*), East Coast fever/theileriosis (*Theileria parva*), babesiosis (*Babesia microti*, *B. ovis*), Lyme borreliosis (*Borrelia burgdorferi*), ehrlichiosis (*Ehrlichia chaffeensis*), Rocky Mountain spotted fever (*Rickettsia rickettsii*), boutonneuse fever (*Rickettsia parkeri*), Queensland tick typhus (*Rickettsia australis*), Q fever (*Coxiella burnetii*) and numerous arboviruses (Sonenshine, 1993). Disease agents can pass from one tick developmental stage to the other. The infected adult female transmits to eggs (transovarial transmission) and to the next generation from larvae to the nymph to the adult (transtadial transmission). Due to their efficiency as vectors of a wide variety of pathogens, broad vertebrate host range and worldwide distribution, ixodid ticks are recognized as potential vectors of a number of pathogens of bio- or agro-terrorism significance including Crimean-Congo hemorrhagic fever virus, *R. rickettsii* and the tick-borne encephalitis complex of flaviviruses (Central European tick-borne encephalitis, Far Eastern tick-borne encephalitis, Russian spring and summer encephalitis, Kyasanur forest disease and Omsk hemorrhagic fever), *C. burnetii* (Q fever) and *Francisella tularensis* (tularemia) (www.cdc.gov/od/sap). Infectious agents of domestic animals and wildlife such as *E. ruminantium*, the causative agent of heartwater in cattle, also represent potential zoonotic disease threats that may be vectored by ticks (Daszak *et al.*, 2000).

2 Tick salivary glands

The tick's multifunctional salivary gland is vital to its biological success, and also plays a critical role in transmitting disease-causing agents to its hosts (Sauer *et al.*, 1995). As tick blood-feeding progresses, the rate of salivary fluid secretion increases greatly, enabling the tick to concentrate the blood meal by returning excess water and ions to the host via saliva (Sauer *et al.*, 2000). Salivary secretion in feeding ticks is under neuronal control. Dopamine is the neurotransmitter at the neuro-effector junction that regulates fluid secretion via adenylate cyclase and an increase in cAMP (Sauer *et al.*, 2000). Dopamine also affects the release of arachidonic acid that is subsequently converted to prostaglandins (PGs). PGs are secreted at extremely high levels into tick saliva for export to the host where they impact the host's physiology. The exact role(s) that PGs play in tick feeding is not known, but they exhibit vasodilatory, anti-haemostatic, immunosuppressive and anti-inflammatory effects in mammals (Oliveira *et al.*, 2011; Sa-Nunes *et al.*, 2007). Prostaglandin E_2 (PGE_2) has an autocrine or paracrine role within the salivary gland itself where it interacts with a PGE_2 receptor to induce

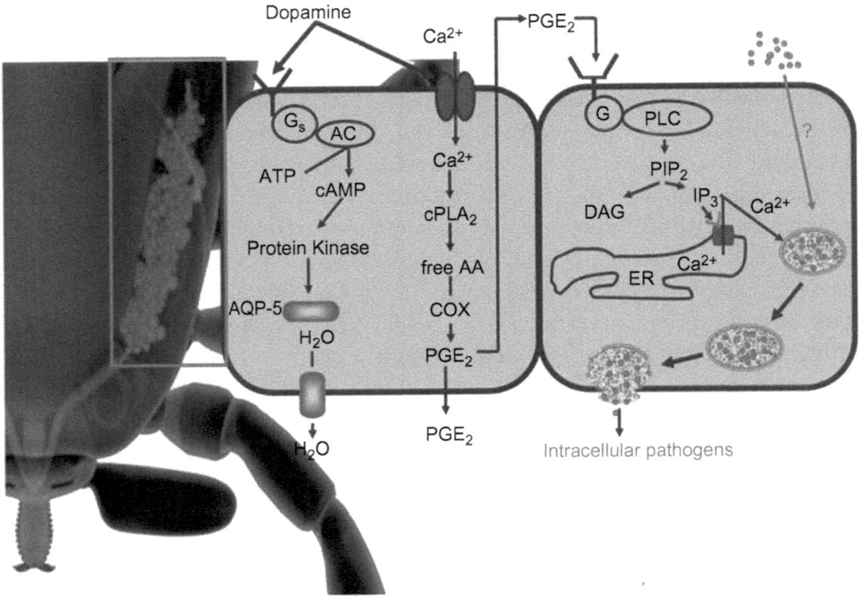

FIG. 1 The known and proposed cascade of reactions controlling prostaglandin E_2 synthesis and tick salivary secretions (Bowman and Sauer, 2004). Dopamine released from nerves stimulate fluid secretion through an increase in cAMP and activation of a water channel protein. Dopamine also stimulates the influx of Ca^{2+}, activating an intracellular PLA_2 and subsequently releasing arachidonic acid (AA) which is converted to PGE_2 by a cyclooxygenase pathway. PGE_2 is either secreted into the saliva or functions as a local hormone by interacting with PGE_2 receptors to increase IP3. IP3 mobilizes intracellular Ca^{2+} which is critical for regulating exocytosis of salivary proteins. Tick-transmitted pathogens are thought to exploit transcytotic pathways, that is they are taken in by endocytosis and released by exocytosis (modified from Bowman and Sauer, 2004). (See Color Plate 8 at the back of the book.)

exocytosis of bioactive proteins via a phosphoinositide signalling pathway and an increase in cellular Ca^{2+} (Fig. 1).

Control of secretion, identification of secretory products and the function of PGE_2 in tick salivary gland physiology have been studied in some detail (Bowman and Sauer, 2004; Sauer *et al.*, 2000). However, surprisingly little is known about the physiology of vector-borne infectious agents trafficking within ticks and particularly salivary glands. Lyme disease spirochetes are thought to exploit transcytotic pathways, that is they are taken in by endocytosis and released by exocytosis (Kurtti *et al.*, 1994). Much of the recent focus on the prevention of arthropod-borne diseases has centered on developing effective vaccines based on pathogen surface antigenic points. However, to more efficiently block bacterial transmission from an arthropod vector, we should thoroughly dissect pathways involved in the survival (entry, multiplication and exit) of intracellular pathogens within salivary cells as well as salivary determinants that help in bacterial secretion. Events which take place in the tick are of importance with respect to pathogen survival within the vector and transmission to vertebrate hosts. Unfortunately, the portion of the life cycle which tick-borne pathogens spend inside the tick (exclusively salivary gland) is an enigma and has barely been investigated. Hypothetically, transmission of pathogens from a tick to its vertebrate host could occur by salivation, regurgitation, defecation or by host ingestion of infected ticks. The feeding habits and central role of salivary secretions in enabling ticks to remain on a host for relatively long periods of time (Kaufman, 1989) strongly suggest that salivation may be an important route of parasite transmission to vertebrate hosts. The large amount of fluid secreted by the salivary glands, facilitating concentration of the blood meal, is well documented (Kaufman, 1989). *A. marginale*, an intracellular pathogen, has been localized in various ixodid tick tissues including midgut epithelium, muscle cell lining, the visceral midgut, haemolymph and salivary glands (Kocan *et al.*, 2004).

Tick saliva is the key to a successful blood meal, which depends upon a fine balance between the action of tick salivary factors and host responses. Tick salivary secretions play a key role in the modulation of host hemostatic and immunological responses (Sauer *et al.*, 2000). Several saliva factors (protein and non-protein) with a diverse range of pharmacological properties have been characterized in tick saliva (for the most comprehensive review see Francischetti *et al.*, 2009). Tick saliva has a plethora of pharmacologically active components that probably also aid in the transmission of tick-borne pathogens to a susceptible host. These include protein components with anti-complement, anti-inflammatory and antihistaminic activities as well as a variety of other factors that inhibit the host immune response (Bowman *et al.*, 1997; Francischetti *et al.*, 2009; Ribeiro and Francischetti, 2003; Sauer *et al.*, 2000). Key determinants of pathogen survival (*Ehrlichia* or *Rickettsia*) in ixodid vectors are unknown. Infection is only transmitted to vertebrate hosts after the ticks are allowed to feed. Despite the central role of the salivary glands in vector competence, our understanding of how the vector responds to parasite infection is very limited. Much of what is known about arthropod-borne parasite multiplication is based on studies performed in cell lines with less physiological information from tick tissues. While the former studies are quite useful, they do not provide a complete picture of the actual infection process within intact salivary glands. Previous data suggest that our basic understanding of tick-borne pathogen infection may be biased by the cell line chosen to study bacterial–vector interactions. It is clear that more effort is required to understand tick–pathogen interactions not only in cell lines but also in infected ticks.

3 Tick genomic resources

The genome of the black-legged deer tick (*Ixodes scapularis*) is the first tick genome sequence available (Hill and Wikel, 2005). Expressed sequence tags (ESTs) are available for a wider range of species, and have become invaluable resources in the area of gene expression analysis, peptide identification and characterization of proteomes, particularly in the absence of complete genome sequence information. As 'tags' to identify genes, ESTs are primarily single pass complementary DNA (cDNA) sequences derived from transcribed mRNAs (transcripts). High-throughput transcriptomic studies of tick salivary glands have been completed for several tick species including *I. scapularis* (Ribeiro *et al.*, 2006; Valenzuela *et al.*, 2002), *Ixodes pacificus* (Francischetti *et al.*, 2005a), *Dermacentor andersoni* (Alarcon-Chaidez *et al.*, 2007), *Amblyomma maculatum* (Karim *et al.*, 2011), *Amblyomma variegatum* (Nene *et al.*, 2002; Ribeiro *et al.*, 2011), *Ixodes ricinus* (Chmelar *et al.*, 2008) and *R. microplus* (Guerrero *et al.*, 2010; Wang *et al.*, 2007). A total of ~297,394 ESTs from a variety of tick species were found in the National Center for Biotechnology Information (NCBI) (as of June 30, 2011). Tick transcriptome and proteome projects in the past 5 years have generated huge datasets that have revealed the complexity of tick saliva, including the presence of many factors with low expression levels. In an attempt to get a global view of salivary proteins, Francischetti *et al.* (2009) assembled EST collections and extracted 3454 ORFs for 13 tick species spanning the Ixodidae. A catalogue of salivary transcripts of *A. maculatum* revealed 3475 contigs associated with a secretory function (Karim *et al.*, 2011). Insight into the sialomes of ticks provides a much more complex picture than originally thought (see Francischetti *et al.*, 2009): (1) saliva has a great diversity of factors, (2) saliva is full of isoforms belonging to multi-gene families and targeting one protein triggers the expression of other isoforms (compensation effect), (3) <5% of the saliva proteins have a known function and (4) saliva proteins are exploited by disease agents to evade the host immune system. The class of expressed tick proteins predicted to be secreted includes diverse gene families such as protease inhibitors, metalloproteases, immunomodulators, antimicrobial peptides, basic tail and glycine-rich peptides (Francischetti *et al.*, 2009). It has been speculated that the diversity of salivary proteins found in tick saliva may be dictated by the adaptive immunity of their hosts (Francischetti *et al.*, 2009). Over ~6500 coding sequences (Francischetti *et al.*, 2009; Karim *et al.*, 2011) are possibly associated with a blood-feeding role as secreted salivary proteins. Members of over 133 tick protein families catalogued from different tick species await functional analysis. Below, we highlight select gene families likely to be important in prolonged blood feeding.

3.1 UBIQUITOUS PROTEIN DOMAINS

3.1.1 *Protease inhibitor-containing domains*

Kunitz domain-containing proteins. Proteins containing single or multiple Kunitz domains have been functionally characterized in ticks (Francischetti *et al.*, 2004; Monteiro *et al.*, 2005, 2008; Nazareth *et al.*, 2006). Kunitz family proteins are diverse, containing 1–6 Kunitz domains. Gene duplication has been proposed in the expansion of protein families associated with blood feeding (Mans and Neitz, 2004; Mans *et al.*, 2002).

TIL domain-containing proteins. Members of the TIL domain-containing protein family have been found ubiquitously in blood-feeding insect and tick sialomes. These polypeptides may also perform antimicrobial functions (Li *et al.*, 2006).

Cystatins. Cystatins are cysteine protease inhibitors of nearly 100 amino acids in length. Cystatins have previously been found in both hard and soft tick sialotranscriptomes. Two salivary cystatins from *I. scapularis* have been functionally characterized as inhibitors of cathepsins L and S; as a result, they inhibit inflammation, suppress dendritic cell maturation and can serve as vaccine targets (Kotsyfakis *et al.*, 2006, 2007).

Serpins. Serpins are a ubiquitous protein family associated with the function of serine protease inhibition. Tick serpins have been proposed as vaccine targets (Imamura *et al.*, 2005; Mulenga *et al.*, 2001).

Kazal domain. The Kazal domain is also associated with serine protease inhibitors and antimicrobial activity (Kanost, 1999; Rawlings *et al.*, 2004).

Carboxypeptidase inhibitors. To date this protein family is found only in metastriate ticks and may affect fibrinolysis (a key step in maintaining an uninterrupted blood supply) by inhibiting plasma carboxypeptidase B, also known as thrombin-activatable fibrinolysis (Arolas *et al.*, 2005).

3.1.2 *Enzymes*

Proteases. Metalloproteases of the reprolysin family have been commonly found in tick sialotranscriptomes (Francischetti *et al.*, 2009; Harnnoi *et al.*, 2007; Mans *et al.*, 2008) and have been associated with fibrinogenolytic and anti-angiogenic activities found in saliva (Francischetti *et al.*, 2003, 2005b). Serine proteases are also commonly found expressed in tick sialotranscriptome but rarely characterized for their substrate specificity. Aspartyl peptidases are found commonly in midgut transcriptomes of ticks (Alim *et al.*, 2007, 2008a,b, 2009; Horn *et al.*, 2009; Sojka *et al.*, 2007).

3.2 SECRETED IMMUNITY-RELATED PROTEINS

Lipocalins. The lipocalin family is widespread in nature and is abundantly expressed in tick sialotranscriptomes as members of a large gene family. In ticks, lipocalins have been associated as binders of histamine and serotonin, and also with anti-complement activity, immunoglobulin binding and with toxic properties (Francischetti *et al.*, 2009). It is quite interesting that lipocalins from *Amblyomma* species are at best only 40% identical to other tick lipocalins (Karim *et al.*, 2011).

Antimicrobial peptides. Tick salivary transcriptomes have revealed abundant expression of antimicrobial peptides that might prevent microbial growth in the imbibed blood and prevent contamination of the feeding cavity with microbes.

Antimicrobial peptides of the defensin, hebeain/microplusin/Ricinusin and lysozyme families have been indentified in different tick species (Kopacek *et al.*, 2010).

3.3 TICK-SPECIFIC PROTEINS

The sialotranscriptome of ticks revealed several tick-specific protein families including (1) glycine-rich proteins, (2) Ixodegrins, (3) Disintegrins, (4) Salp15, (5) Basic tail 18.3-kDa superfamily, (6) 23-kDa family, (7) 8.9-kDa family, (8) Ixostatin, (9) one of each family and (10) Novel family 40–33 (Karim *et al.*, 2011). Listed tick-specific proteins have diverse functions associated with them. The majority of these families await functional characterization.

3.4 METASTRIATE-SPECIFIC PROTEINS

Evasins. Evasins are chemokine-binding proteins identified only in metastriate ticks (Deruaz *et al.*, 2008; Frauenschuh *et al.*, 2007). Within *Amblyomma*, analysis revealed at least 18 different genes coding for members of the Evasin family although this family is not particularly highly expressed (Karim *et al.*, 2011).

DAP-36 immunosuppressant. A 36-kDa salivary protein has been identified as an immunosuppressant from *D. andersoni* (Bergman *et al.*, 2000, 1998). It is found to be divergent and widespread in metastriate ticks (Francischetti *et al.*, 2009). Several members of this family have been found in *A. maculatum* (Karim *et al.*, 2011) and await functional characterization.

Metastriate 13-kDa family. The tick sialotranscriptome projects also revealed 10–11 kDa proteins with a mature molecular weight of (Francischetti *et al.*, 2009). The function of this family is unknown.

3.5 DEORPHANIZED METASTRIATE-SPECIFIC PROTEIN FAMILIES

The sialotranscriptome of *A. maculatum* revealed 12 novel protein families that match previously orphan proteins known only from other metastriate ticks (Karim *et al.*, 2011).

3.6 SPECIES-SPECIFIC PROTEINS

A total of 54 families significantly match solely to previously described *Amblyomma* proteins. With additional tick genome and transcriptome sequencing, it is expected that many of these proteins will be deorphanized.

The newly identified tick gene family data set (~133 gene families containing ~6500 coding sequences) should serve as a starting point to develop an RNAi screening platform, as discussed later in this chapter.

4 RNA interference (RNAi)

Gene silencing through RNAi was first observed by plant biologists in the late 1980s, but its molecular mechanism remained unclear until the late 1990s, when an elegant study in the nematode *Caenorhabditis elegans* showed that RNAi is dependent on the presence of double-stranded (ds)RNA matching the target gene (Fire *et al.*, 1998; Reinhart *et al.*, 2000). This sequence-specific post-transcriptional gene-silencing mechanism is conserved in a range of organisms including plants, *Neurospora*, *C. elegans*, *Drosophila*, mosquitoes, ticks and mammals (Cogoni and Macino, 2000; Mello and Conte, 2004; Ramakrishnan *et al.*, 2005). This discovery heralded a revolution in biology (especially vector biology), because suddenly a reverse genetic tool was available that allowed fast analysis of gene function which was not restricted to model organisms, such as *Drosophila* (Belles, 2010; de la Fuente *et al.*, 2007; Meister and Tuschl, 2004; Mello and Conte, 2004; Tomoyasu *et al.*, 2008).

In a relatively short time, RNAi has rapidly progressed to become a key experimental tool. Inhibition of protein secretion, decrease in transcript level and reduction in encoded proteins have been demonstrated in tick salivary glands using *in vitro* and *in vivo* dsRNA-mediated interference (Aljamali *et al.*, 2003; de la Fuente *et al.*, 2007; Karim *et al.*, 2004a,b, 2005, 2008, 2010; Kotsyfakis *et al.*, 2007; Narasimahan *et al.*, 2004). A complete list of functionally characterized tick target genes using RNAi is assembled in Table 1. RNAi-mediated gene silencing has been characterized as two partially overlapping and one independent pathways, the RNAi pathway, miRNA pathway and piRNA pathway. The RNAi pathway is triggered by exogenous or endogenous dsRNAs that are recognized by Dicer RNase III proteins, which 'chop' long dsRNA into small interfering RNAs (siRNAs) of 21–23 nt in length (Ketting, 2011). The siRNAs are recruited into RNA-induced silencing complexes (RISCs) and dsRNA-binding motif proteins (dsRBMs) that facilitate siRNAs to be joined correctly into silencing complexes (Ketting, 2011; Liu *et al.*, 2003; Tabara *et al.*, 2002). The RISCs find target mRNA using siRNAs as a cue and chop it into pieces. Argonaute (Ago) family proteins are the main components of silencing complexes, facilitating target recognition and silencing (Kurscheid *et al.*, 2009; Liu *et al.*, 2004; Meister and Tuschl, 2004; Song *et al.*, 2004).

The exceptional gene specificity and effectiveness of the RNAi mechanism to modulate gene expression has not only provided a powerful tool to study the gene function in basic research but may also revolutionize targeting vector physiology to control vector-borne infections. The popularity of RNAi has made it a mandatory lab tool in almost all research laboratories investigating tick physiology, tick–pathogen interactions (pathogen survival in tick and transmission to the host) and vector competence because of its demonstrated efficient, robust and reproducible post-transcriptional gene-silencing effects in arthropod vectors (Cogoni and Macino, 2000).

TABLE 1

Summary of RNAi as a method of functional genomics to study tick physiology

Target gene	Amount of dsRNA	Phenotype
Amblyomma americanum (lone-star tick)		
Insulin-like growth factor binding protein (AamIGFBP-rP1)	2000–3000ng	Reduction in blood meal size, failure to lay eggs, and higher mortality (Mulenga and Khumthong, 2010b)
Insulin-like growth factor binding protein AamIGFBP-rP6(L) and AamIGFBP-rP6(S)	2000–3000ng	Reduction in blood meal size (Mulenga and Khumthong, 2010b)
Serine-protease Inhibitor 6 (AamS6)	–	No observed phenotype (Chalaire *et al.*, 2011)
CD147 receptor/extracellular matrix metalloprotease inducer (EMMPRIN)	750–3000ng	Reduction in blood meal sizes and failure to spontaneously detach from their host (Mulenga and Khumthong, 2010a)
Cystatin	500, 25,000ng salivary glands	Decreased tick feeding and lower tick weight (Karim *et al.*, 2010)
Histamine-binding protein (HBP)	1000, 8000ng salivary glands	Reduction of histamine-binding activity in salivary glands and aberrant tick feeding pattern (Aljamali *et al.*, 2002, 2003)
Interphase cytoplasm foci protein 45 (2G7)	2.5–250 $\times 10^9$ molecules	Reduction in tick weight, oviposition and egg fertility (de la Fuente *et al.*, 2010)
Macrophage migration inhibitory factor (MIF)	–	No physiological phenotype apparent (Bowen *et al.*, 2010)
Neuronal isoform munc18-1 (nSec1)	15,000ng for isolated salivary glands, 450ng injected	Reduction in anticoagulant protein release by salivary glands and aberrant tick feeding (Karim *et al.*, 2004a)
Organic anion transporter polypeptide (OATP)	3000ng	Smaller blood meals, blocked spontaneous detachment and laid fewer eggs (Mulenga *et al.*, 2008)
Putative threonyl-tRNA synthetase (2C9)	2.5–250 $\times 10^9$ molecules	Reduction in tick weight, oviposition and egg fertility (de la Fuente *et al.*, 2010)
Subolesin (4D8)	2–60 $\times 10^{11}$ molecules	Reduction in tick survival, weight and oviposition (de la Fuente *et al.*, 2006a)

Synaptobrevin	500ng	Inhibition of PGE2-stimulated anticoagulant protein secretion by isolated salivary glands (Karim *et al.*, 2005)
60S ribosomal protein (2D10)	2.5–250×10^9 molecules	Reduction in tick weight, oviposition and egg fertility (de la Fuente *et al.*, 2010)
60S ribosomal protein (2B7)	2.5–250×10^9 molecules	Reduction in tick weight, oviposition and egg fertility (de la Fuente *et al.*, 2010)
Dermacentor marginatus (ornate sheep tick)		
Subolesin (4D8)	2–60×10^{11} molecules	Reduction in tick survival, weight and oviposition (de la Fuente *et al.*, 2006a)
Dermacentor variabilis (American dog tick)		
Glutathione *S*-transferase (GST)	2–20×10^{10} molecules	Salivary glands not infected with *Anaplasma marginale* (Kocan *et al.*, 2009)
Selenoprotein M (Sel M)	2–20×10^{10} molecules	Salivary glands not infected with *Anaplasma marginale* (Kocan *et al.*, 2009)
Subolesin (4D8)	2–60×10^{11} molecules; 2–20×10^{10} molecules	Reduction in tick survival, weight and oviposition. Decreased reproductive performance and no viable offspring; salivary glands not infected with *Anaplasma marginale*. (de la Fuente *et al.*, 2006b,c, Kocan *et al.*, 2009)
Varisin	–	Antimicrobial activity of the hemolymph plasma was reduced two- to fourfold (Hynes *et al.*, 2008)
Vitellogenin receptor (VgR)	500ng	Decreased oviposition and brown eggs develop (Mitchell *et al.*, 2007)
H+ transporting lysosomal vacuolar proton pump (vATPase)	2–20×10^{10} molecules	Salivary glands not infected with *Anaplasma marginale* (Kocan *et al.*, 2009)
Haemaphysalis longicornis (New Zealand cattle tick)		
Actin	1–10×10^9 molecules	Inhibition of tick feeding and reduced tick weight (de la Fuente *et al.*, 2005, Narasimhan *et al.*, 2004)
Class B scavenger receptor (HlSRB)	1000ng	Reduction in tick weight (Aung *et al.*, 2011)
Cubilin-related serine proteinase (HlSP)	1000ng	Reduction in tick weight (Miyoshi *et al.*, 2004)
Cytosolic leucine aminopeptidase (HlLAP)	1000ng	Reduction in oviposition (Hatta *et al.*, 2007)

(*continues*)

TABLE 1 (*Continued*)

Target gene	Amount of dsRNA	Phenotype
Follistatin-related protein (FRP)	1000ng	Reduction in tick oviposition (Zhou *et al.*, 2006)
Hl86	1000ng	Reduced tick body weight after engorgement (Liao *et al.*, 2007)
HlYkt6	–	Lower engorged body weight, higher mortality of replete ticks, suppression of saliva secretion and lower anticoagulant activity in saliva (Gong *et al.*, 2009)
Longepsin	2500ng	Reduction in longepsin mRNA and protein levels (Boldbaatar *et al.*, 2006)
Lysine-ketoglutarate reductase/saccharopine dehydrogenase (LKR/SDH)	2500ng	Strongly decreases egg viability and reduction in tick body weight after engorgement (Battur *et al.*, 2009)
Midgut-specific asparaginyl endopeptidases/legumain (HlLgm)	1000ng	Higher tick mortality, lower tick body weight, damage to midgut tissues, significant delay to onset of oviposition and reduced number of eggs and egg inviability (Alim *et al.*, 2009)
Midgut-specific asparaginyl endopeptidases/legumain (HlLgm2)	1000ng	Higher tick mortality, lower tick body weight, damage to midgut tissues, significant delay to onset of oviposition and reduced number of eggs and egg inviability (Alim *et al.*, 2009)
Protein disulfide isomerase 1 (HlPDI-1)	1000ng	Impaired tick blood feeding and decreased oviposition (Liao *et al.*, 2008)
Protein disulfide isomerase 2 (HlPDI-2)	1000ng	Decreased tick viability (Liao *et al.*, 2008)
Protein disulfide isomerase 3 (HlPDI-3)	1000ng	No significant phenotype unless dually injected with dsRNA corresponding to HlPDI-1 which had a synergistic effect on tick viability (Liao *et al.*, 2008)
Ribosomal protein P0 (HlP0)	–	Significantly lower body weight, lower engorgement weight and higher mortality (Gong *et al.*, 2008)
Valosin-containing protein (HlVCP)	2500ng	Lower tick body weight (Boldbaatar *et al.*, 2007)

Vitellogenin-1 (HlVg-1)	1000ng	Lower tick body weight, egg weight and higher mortality (Boldbaatar *et al*., 2010)
Vitellogenin-2 (HlVg-2)	1000ng	Lower tick body weight, egg weight and higher mortality (Boldbaatar *et al*., 2010)
Vitellogenin-3 (HlVg-3)	1000ng	Lower tick body weight, egg weight and higher mortality (Boldbaatar *et al*., 2010)
Vitellogenin receptor (HlVgR)	1000ng	Eggs did not fully mature and the *Babesia* parasite DNA was not detected (Boldbaatar *et al*., 2008)
Longistatin	1000ng	Lower tick engorgement weight (Anisuzzaman *et al*., 2011)
Ixodes ricinus (Common sheep tick)		
Ferritin (FER1)	1500ng	Lower tick weight after engorgement and reduced hatching rate (Hajdusek *et al*., 2009)
Ferritin (FER2)	1500ng	Lower tick weight after engorgement, reduced hatching rate and impaired ability to feed (Hajdusek *et al*., 2009)
Iron regulatory protein (IRP1)	1500ng	Lower tick weight after engorgement and reduced hatching rate (Hajdusek *et al*., 2009)
Putative metalloprotease 1 (Metis 1)	650ng	Higher tick mortality, reduced engorgement and oviposition rate (Decrem *et al*., 2008)
Putative metalloprotease 2 (Metis 2)	650ng	Higher tick mortality, reduced engorgement and oviposition rate (Decrem *et al*., 2008)
Cathepsin L (IrCL1)	–	Reduced tick engorgement (Franta *et al*., 2011)
Ixodes scapularis (black-legged/deer tick)		
Actin	1000ng	– (Karim *et al*., 2008)
Calreticulin	1000ng	– (Karim *et al*., 2008)
Histamine-release factor (tHRF)	4×10^9 molecules	Impaired tick feeding and decreased *B. burgdorferi* burden in mice (Dai *et al*., 2010)
Antifreeze glycoprotein (IAFGP)	4.2×10^9 molecules	Decreased cold tolerance (Neelakanta *et al*., 2010)
Intracellular cystatin	1000ng	– (Karim *et al*., 2008)

(*continues*)

TABLE 1 (*Continued*)

Target gene	Amount of dsRNA	Phenotype
5′ Nucleotidase (4D8)	1–10 $\times 10^9$ molecules	Lower tick survival, weight and oviposition (de la Fuente *et al.*, 2005)
Putative phospholipase A2 (PLA2)	1000 ng	– (Karim *et al.*, 2008)
Salivary protein 14 (Salp14)	5 $\times 10^9$ molecules	Reduction in anticoagulant activity in salivary glands and lower tick weight (Narasimhan *et al.*, 2004)
Salivary protein 14 (Salp14)	5 $\times 10^7$ molecules	Reduction in Salp14 mRNA levels but no effect on tick feeding (Pedra *et al.*, 2006)
Sialostatin L2	–	Immediate tick–host rejection, mortality and reduced engorged body weight (Kotsyfakis *et al.*, 2006)
Semliki forest virus (SFV) replicon	–	Suppression of RNAi induced by SFV replicon (Garcia *et al.*, 2006)
Subolesin (4D8)	2–60 $\times 10^{11}$ molecules	Reduction in tick weight and oviposition (de la Fuente *et al.*, 2005, 2006a,b,c)
Subolesin (4D8)	–	Reduction in Subolesin mRNA levels (Blouin *et al.*, unpublished data)
Syntaxin 5	1000 ng	– (Karim *et al.*, 2010)
Rhipicephalus annulatus (brown ear tick)		
Elongation factor 1-alpha (EF-1α)	1.5–25 $\times 10^{10}$ molecules	Reduces oviposition and hatching (Almazán *et al.*, 2010)
GST	1.5–25 $\times 10^{10}$ molecules	Impaired attachment (Almazán *et al.*, 2010)
Selenoprotein W (SEL)	1.5–25 $\times 10^{10}$ molecules	Impaired attachment (Almazán *et al.*, 2010)
Subolesin	1.5–25 $\times 10^{10}$ molecules	Higher mortality and lower oviposition (Almazán *et al.*, 2010)
Ubiquitin (UBQ)	1.5–25 $\times 10^{10}$ molecules	Higher mortality and lower oviposition (Almazán *et al.*, 2010)
Rhipicephalus evertsi evertsi (band-legged tick)		
ReeATAQ	2.5–3.5 $\times 10^{11}$ molecules	No significant phenotype (Nijhof *et al.*, 2010)
Ree86	2.5–3.5 $\times 10^{11}$ molecules	No significant phenotype (Nijhof *et al.*, 2010)
Rhipicephalus haemaphysaloides		
Rhipilin-1	1000 ng	Fifty-three percent decrease in tick attachment rate (Gao *et al.*, 2011)

Rhipicephalus microplus (*Boophilus microplus*) (cattle tick)		
Bm86	$3–6\times10^{11}$ molecules	Systemic RNAi in tick progeny (Nijhof *et al.*, 2007)
Bm91	$3–6\times10^{11}$ molecules	Systemic RNAi in tick progeny (Nijhof *et al.*, 2007)
Elongation factor 1-alpha (EF-1α)	$1.5–25\times10^{10}$ molecules	Higher mortality and lower oviposition (Almazán *et al.*, 2010)
Glycogen synthase kinase-3 (GSK-3)	–	Reduces oviposition and hatching (Fabres *et al.*, 2010)
GST	$1.5–25\times10^{10}$ molecules	Impaired attachment (Almazán *et al.*, 2010)
Selenoprotein W (SEL)	$1.5–25\times10^{10}$ molecules	Impaired attachment (Almazán *et al.*, 2010)
Subolesin (4D8)	$3–6\times10^{11}$ molecules, $1.5–25\times10^{10}$ molecules	Reduction in tick survival, weight and oviposition. Abnormal egg development. Systemic RNAi on tick progeny, higher mortality and decrease oviposition (Almazán *et al.*, 2010; Nijhof *et al.*, 2007).
Ubiquitin (UBQ)	$1.5–25\times10^{10}$ molecules	Higher mortality and lower oviposition (Almazán *et al.*, 2010)
Ubiquitin-63E	5 μL of 3.33 μM soln	This study demonstrated that off-target effects produced by the injection of dsRNA could be reduced by shortening the size of the dsRNA injected (Lew-Tabor *et al.*, 2011.)
Rhipicephalus sanguineus (brown dog tick)		
Rs86 (homologue of Bm86)	$9–30\times10^{9}$ molecules	Reduction in tick weight and oviposition (de la Fuente *et al.*, 2006c)
Subolesin (4D8)	$2–60\times10^{11}$ molecules; $9–30\times10^{9}$ molecules	Reduction in tick survival, weight and oviposition (de la Fuente *et al.*, 2006a,c)

RNAi has emerged as a routine lab tool to discover functions of unknown genes (~95% of tick genes deposited in public database). For *in vivo* application in ticks, the most convenient way to achieve gene silencing is by injection of dsRNA into the tick hemocoel. This approach, in which gene silencing is achieved in tick tissues after endocytosis of dsRNA, is termed 'systemic RNAi' and has led to successful gene depletion of the target gene/s in reported tick species (Table 1). The proteins involved in the systemic spread of siRNA and dsRNA uptake in ticks have been identified in *R. microplus* and *I. scapularis* (Ago-2, Rsd-3 and Epn-1) (de la Fuente *et al.*, 2007; Kurscheid *et al.*, 2009). Next generation sequencing has paved the way for increased tick genomic resources and it may be feasible to identify more putative tick homologues of essential RNAi pathway-associated proteins to gain insight into the tick RNAi mechanism. It may also shed light on the conservation of the RNAi mechanism within different tick species. A better understanding of RNAi mechanisms employed by different tick species for gene silencing will facilitate RNAi screen reagents for high-throughput gene function studies. An elegant study revealed the key RNAi-associated proteins and loss-of-function in knockdown mutants suggest a functional RNAi pathway in *R. microplus* (Kurscheid *et al.*, 2009). The investigation also indicates that a blood-feeding arthropod RNAi pathway may differ from that in other arthropods like insects. In *D. melanogaster*, siRNAs and miRNAs are generated by distinct Dicer enzymes (Lee *et al.*, 2004) while a single Dicer protein is in *C. elegans*, humans and ticks (Kurscheid *et al.*, 2009). To gain in-depth knowledge of the proposed RNAi mechanism in ticks, we recommend reading Kurscheid *et al.* (2009).

RNAi offers a method of specifically inactivating expression (gene silencing) of the corresponding gene and thereby serves as a method for investigating gene function (Carthew, 2001). The use of RNAi has become a powerful experimental tool for studies involving tick–host–pathogen interaction but its utility has been severely limited by the tedious and often traumatic nature of delivering dsRNA to immature stages of ticks by microinjections or capillary feeding (Pal *et al.*, 2004; Soares *et al.*, 2005). It is generally assumed that RNAi will always occur once dsRNA is delivered inside the tick cell. While RNAi is greatly facilitating studies to better understand specific gene function, the major challenges in using dsRNA among non-model organisms like ticks are (1) selection of a target gene, (2) delivery of long dsRNA (200–500 bp), (3) determination of the volume and concentration of dsRNA required for extensive gene knockdown, (4) prolonged attachment and (5) siRNA dilution factor. The volume and concentration of dsRNA required to silence genes varies from species to species (e.g. *I. scapularis* requires a low volume as compared to *Amblyomma americanum*; Karim *et al.*, 2004a,b, 2005). Before starting an RNAi experiment, the gene expression of a target gene can be a valuable tool to select an appropriate dsRNA concentration (abundantly expressed transcript may require more dsRNA) for complete transcript depletion (Fig. 2). The following methods have been proven successful in exogenous delivery of long dsRNA in different developmental stages of tick species.

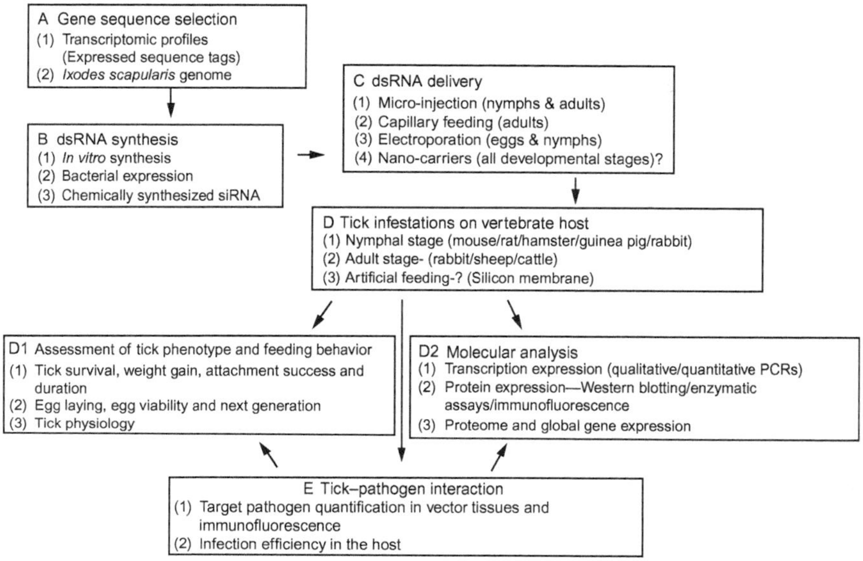

FIG. 2 Typical steps in tick RNAi experiments. (A) RNAi experiments start with the selection of a target gene sequence; (B) dsRNA is typically synthesized by (1) *in vitro* transcription using off-the-shelf kits, (2) bacterial expression of dsRNA using the L4440 vector and (3) siRNA using computational biology methods; (C) dsRNA or siRNA is typically delivered in the tick body by (1) microinjections, (2) capillary feeding, (3) electroporation and (4) nano-carriers (potential future application); (D) ticks are infested on an appropriate vertebrate host after the delivery of dsRNA; (D1) post-tick attachment, tick survival, weight gain, attachment success, oviposit and tick physiology is assessed; (D2) transcription expression (qualitative or quantitative RT-PCRs) and protein expression is determined to test the gene disruption and (E) RNAi experiments with pathogen-infected ticks typically quantify the pathogen in tick tissues and host by qPCR.

5 Delivery of dsRNA in ticks

5.1 MICROINJECTIONS

Direct injection is a facile and safe method to deliver long dsRNA to the target location in ticks. It has been widely used in multiple tick species and also insect species including mosquitoes, beetles, honey bees, grasshoppers and aphids (Amdam *et al.*, 2003; Blandin *et al.*, 2002; Bucher *et al.*, 2002; Dong and Friedrich, 2005; Dzitoyeva *et al.*, 2001; Jaubert-Possamai *et al.*, 2007; Karim *et al.*, 2004a). Microinjection is routinely used in adult ticks and requires a 24-h incubation time for ticks to heal the site of injection and recover before infesting them on the vertebrate host. Microinjection of dsRNA into unfed nymphs has also been reported (Pal *et al.*, 2004); however, this requires special handling, skills and training to get reproducible results (Karim *et al.*, 2010).

5.2 CAPILLARY FEEDING

Artificial feeding is one example of a non-traumatic method for delivering dsRNA that preserves the integrity of the treated organism. However, the precise amount of dsRNA taken up can be difficult to monitor. Capillary feeding of dsRNA has been applied with greatest effect in *I. scapularis* (Soares *et al.*, 2005). Capillary feeding of the immature tick stage can be used to study the function of specific nymphal tick genes involved in blood feeding and pathogen transmission. A complicating factor in RNAi-silenced nymphal ticks is the undifferentiated sexual dimorphism, which could lead to variable gene silencing and pathogen transmission results.

5.3 ELECTROPORATION

Electroporation is a powerful transfection tool useful for studying gene function. The principle application of electroporation is focused on vertebrate tissue and organisms; recent work demonstrated the efficacy of this tool for delivering dsRNA into tick eggs and nymphs (Karim *et al.*, 2010). Electroporation permeabilizes the membranes of cells when an electrical current is applied. The long dsRNA is usually delivered in a high ionic strength medium. Optimization of electroporation requires consideration of factors including tick developmental stage, dsRNA dose, electrode design, electrical field and duration (Karim *et al.*, 2010). Electroporation is a useful method for delivering long dsRNA *in vivo* to immature tick stages where injections and capillary feeding are not feasible. It is a technique that is also being investigated for dsRNA delivery in tick eggs. Ease of electroporation for gene silencing in tick nymphal or egg stage would make RNAi more widely available to tick biologists and enhance future RNAi applications in tick–host–pathogen interaction and high-throughput tick functional genomic research.

6 RNAi targets in tick salivary glands

6.1 RNAI TO ELUCIDATE THE MECHANISM OF TICK SALIVARY SECRETION

Tick saliva contains numerous bioactive proteins as documented from biochemical analysis of secretagogue-induced saliva (Bowman *et al.*, 1997; Sauer *et al.*, 1995). Most of the proteins in tick saliva are thought to be released from the vesicle contents of the membrane-bound granular material observed in salivary gland acini type II and III (Binnington, 1978). The mechanism for intracellular trafficking of secretory vesicles and exocytosis is complex. Highly conserved SNARE (soluble *N*-ethylmaleimide-sensitive factor attachment receptors) complex proteins have been identified in all neuronal and non-neuronal secretory cell studies (Rothman, 1994). SNARE terminology is based upon the requirement of

N-ethylmaleimide-sensitive fusion protein (NSF) and soluble NSF attachment protein (SNAP) involvement in the process. SNARE proteins associated with vesicles are referred to as v-SNARE and those associated with the plasma membrane are t-SNARE. NSF uses energy from ATP hydrolysis to dissociate SNARE complexes after membrane fusion, enabling SNARE proteins to be recycled for subsequent rounds of fusion (May *et al.*, 2001). GTP-binding proteins (e.g. Rab3A) are localized to the cytosolic face of specific intracellular secretory vesicle membranes where they are thought to function in vesicle trafficking. Additional components exist that inhibit fusion of vesicle and plasma membranes until the cell receives an appropriate signal, typically a rise in intracellular Ca^{2+} (Karim *et al.*, 2002). A major Ca^{2+} signalling protein is synaptotagmin (Geppert *et al.*, 1994). Proteins in the salivary glands of a partially fed female lone star tick cross-react individually with antibodies to SNARE complex proteins (Karim *et al.*, 2002). Significantly, antibodies to these SNARE complex proteins inhibit PGE_2-stimulated secretion of anticoagulant protein in permeabilized tick salivary glands (Karim *et al.*, 2002).

We have been using RNAi to elucidate the function of SNARE proteins in tick salivary secretions (Fig. 3). In response to 'knock down' of the v-SNARE protein Synaptobrevin, PGE_2-stimulated protein secretion was inhibited in isolated *A. americanum* salivary glands (Karim *et al.*, 2004b). The *nsf*, a unique and single SNARE gene in ticks, is highly conserved from yeast to human. It is a member of the ATPase-associated activity superfamily. Tick *nsf* gene was disrupted in unfed adult female *Dermacentor variabilis* to assess its functional role in tick feeding and salivary secretion (Karim *et al.*, 2008, 2010). The transcriptional expression of the NSF gene was completely depleted in partially fed tick tissues (Fig. 3A). Histochemical analysis revealed the inhibition of secretory vesicle fusion with the plasma membrane and impaired tick feeding on the host in NSF-silenced ticks (Fig. 3B and C). Depletion of the NSF gene transcript also impaired weight gain and oviposition (Fig. 3D). The results strongly suggest that NSF plays a key role in Ca^{2+}-regulated protein secretion in tick salivary glands. The control and mechanism of vesicle transport are likely important for pathogen transmission. Lyme disease spirochetes are believed to move across cells by a transcytotic mechanism, entering by endocytosis and exiting by exocytosis (Hechemy *et al.*, 1992; Kurtti *et al.*, 1994). It seems likely that pathogen delivery to the host may be facilitated by the tick's mechanisms of transport.

6.1.1 *SNARE complex proteins*

nSec1. nSec1 plays a role in promoting fusion of carrier vesicles and target compartments and is required for exocytosis. dsRNA-mediated knockdown of nSec1 resulted in a decrease in anticoagulant protein released by salivary glands, which impaired feeding (Karim *et al.*, 2004b). This study also demonstrated it was possible to use dsRNA to deplete nSec1 transcriptional expression *in vitro* in isolated salivary glands (Karim *et al.*, 2004a).

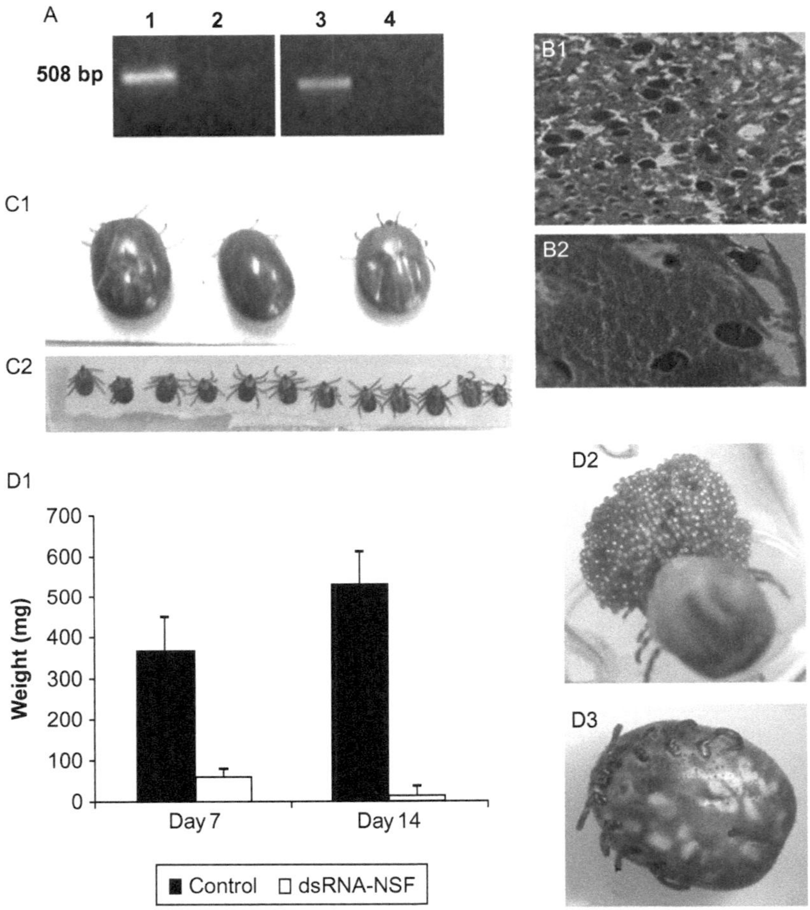

FIG. 3 *In vivo* RNAi in ticks. The *N*-ethylmaleimide-sensitive fusion factor (NSF) dsRNA and an irrelevant GFP-dsRNA delivered by microinjection in unfed female *Dermacentor variabilis*. After microinjections of dsRNA, ticks are left overnight in an incubator and infested on a rabbit. (A) Transcriptional expression of NSF gene in partially fed tick tissues, (1,3) control midgut and salivary gland tissues, (2,4) NSF-dsRNA-injected midgut and salivary gland tissues. (B1 and B2) Thick sections of control and NSF-dsRNA salivary glands stained with dye demonstrating the disruption of secretory vesicles. (C1 and C2) Tick engorgement in control and NSF-dsRNA-treated ticks after 15 days of attachment on rabbit. (D1) average engorgement weight of ticks on day 7 and 14 among control and treated tick groups. (D2 and D3) Oviposition in control and dsRNA-NSF injected ticks. (See Color Plate 9 at the back of the book.)

Ykt6. Ykt6 is another key component of the SNARE complex that participates in vesicle trafficking. Knockdown of Ykt6 by RNAi led to inhibited salivary secretions and lower anticoagulant activity. These data lead the authors to suggest that Ykt6 is involved in exocytosis of salivary proteins and is necessary for tick feeding (Gong *et al.*, 2009).

Synaptobrevin. Synaptobrevin is a v-SNARE protein thought to play a key role in exocytotic protein secretion. RNAi-targeting synaptobrevin disrupted tick feeding by inhibiting feeding and reduced tick engorgement weights (Karim *et al.*, 2004b).

6.2 PROTEASES

Longepsin. The genome of *Haemaphysalis longicornis* encodes a wide variety of proteases that degrade haemoglobin and other proteins in the host blood. Longepsin, which encodes an aspartic protease, seems to be important for blood meal utilization and recombinant longepsin was able to hydrolyse casein substrate. RNAi did not yield a strong phenotype, possibly due to dilution effects of dsRNA (Boldbaatar *et al.*, 2006).

Cytosolic leucine aminopeptidase. Depletion of a *H. longicornis* leucine aminopeptidase resulted in delayed egg laying and reduced oviposition (Hatta *et al.*, 2007). These two effects are presumably a result of disrupting one of the proteases necessary to convert imbibed blood proteins into yolk protein needed for egg development (Hatta *et al.*, 2007).

Metis 1/2. Two putative metalloproteases were identified in *I. ricinus*, and were induced in salivary glands during feeding. Several metalloproteases present in *I. scapularis* saliva have fibrinolytic activity, and may function as anti-haemostatic or anti-angiogenic factors. RNAi-targeting Metis 1 and 2 interfered with the tick's ability to reduce fibrinolysis time. Gene depletion also resulted in reduced tick engorgement rate and egg laying, but not survival rate or feeding time (Decrem *et al.*, 2008).

6.3 PROTEASE INHIBITORS

Cystatins. Cystatins (including sialostatin L and L2) are low molecular mass (13–15kDa) cysteine protease inhibitors identified in a range of arthropods and vertebrates. Cystatins from nematodes are thought to suppress proteases involved in antigen processing in order to evade the host defence system. Naïve ticks exposed to a rabbit previously exposed to cystatin-silenced ticks exhibited significant resistance to feeding, suggesting tick cystatin may play a role similar to that identified in nematodes. RNAi of *I. scapularis* cystatins resulted in an inability to feed in 40% of ticks; feeding was significantly reduced leading to reduced tick engorgement weights in the remaining ticks (Karim *et al.*, 2004b; Kotsyfakis *et al.*, 2007).

Rhipilin-1. Rhipilin-1 belongs to a family of Kunitz-type protease inhibitors with high homology to a tissue factor pathway inhibitor, which inhibits factor Xa and thrombin. Likewise, factor Xa and thrombin have shown that recombinant rhipilin-1 delayed clotting time. RNAi led to a significant decrease in tick attachment rate and decreased engorged body weight (Gao *et al.*, 2011).

Serine-protease inhibitor 6 (AamS6). Ticks are thought to inject serine protease inhibitors (serpins) into the host to combat host defences. AamS6 is strongly expressed in salivary glands in the early feeding phase. Antibodies to tick salivary proteins reacted specifically with rAamS6. RNAi did not produce an observed phenotype, which may be due to an apparently slow turnover rate (Chalaire *et al.*, 2011).

6.4 OTHER PROTEIN CLASSES

CD147 receptor/extracellular matrix metalloprotease inducer (EMMPRIN). CD147 receptors are cell-surface glycoproteins that are involved in intercellular interactions associated with numerous physiological and pathological processes such as extracellular matrix remodelling. CD147 identified in *A. americanum* is up-regulated in response to feeding in the salivary gland, but expression remains constant in the midgut and ovary. RNAi-induced ticks obtained smaller blood meals and failed to spontaneously detach from their host. Many ticks showed symptoms consistent with midgut damage allowing red blood cells to leak into their haemolymph (Mulenga and Khumthong, 2010a). The authors suggest that silencing this receptor blocked progression of tick intermolt growth, which precedes tick engorgement.

Salp14. Salp14 appears to be a member of potential anticoagulants identified in *I. scapularis*. RNAi-targeting salp14 and its paralogs resulted in a 60–80% reduction in anti-factor Xa activity. Engorged body weight was reduced to 50–70% in knockdown ticks (Narasimhan *et al.*, 2004). The authors suggest that while this anticoagulant would represent a potentially attractive tick vaccine candidate, multiple anticoagulants would need to be targeted simultaneously to completely abolish tick feeding.

Ferritin 2. Ticks must manage the excess iron originating from host blood. One of the key proteins involved in iron metabolism is ferritin. A new secreted ferritin was identified from *I. ricinus*. RNAi targeting the secreted ferritin gene (*fer2*) led to loss of expression of *fer1* in the salivary glands and ovaries, which normally follows blood ingestion. This secreted ferritin (*fer2*) seems to function as the primary transporter of non-heme iron between the tick gut and the peripheral tissues. RNAi also decreased tick engorgement weights and hatching rate (Hajdusek *et al.*, 2009).

Selenoprotein M. RNAi of selenoprotein M resulted in ticks with lower infection rates of *A. marginale* (Kocan *et al.*, 2009).

Histamine-binding proteins. Ticks counter the inflammatory response in part through the secretion of histamine-binding proteins. In *A. americanum*, two histamine-binding proteins have been identified that facilitate feeding. RNAi of these two genes led to a reduced histamine-binding ability in isolated salivary glands and aberrant tick-feeding pattern or host response (Aljamali *et al.*, 2003).

Histamine-release factor. Tick saliva contains a variety of proteins that contain anti-inflammatory properties, including histamine-binding proteins.

Given the harmful effects of histamine on tick feeding, it is perhaps surprising that the *I. scapularis* and *D. variabilis* genomes encode a histamine-release factor. The role of this protein during tick feeding is not well understood. RNAi resulted in a reduced tick engorgement rate and decreased *B. burgdorferi* burden in mice, indicating the importance of the histamine-release factor in pathogen transmission (Dai *et al.*, 2010).

Valosin-containing protein. Valosin-containing protein is classified as a member of the AAA protein family; these are thought to act as molecular chaperones in the assembly/disassembly of protein complexes, folding/unfolding proteins, and transport of cellular cargo. Antisera raised against VCP showed strong reactivity in the agranular part of secretory acini salivary glands and midguts. RNAi targeting this gene led to a 42% reduction in engorged tick weight (Boldbaatar *et al.*, 2006).

Longistatin. Ticks are successful in maintaining a blood pool underneath the skin of host, despite strong clotting machinery, through the use of protein factors that inhibit blood clotting. In *H. longicornis*, a salivary gland protein termed longistatin was shown to prevent the cascade of blood clotting. In ticks treated with dsRNA, the ticks failed to form a blood pool necessary for proper feeding, leading to a reduced engorged tick weight (Anisuzzaman *et al.*, 2011).

6.5 RNAI TARGETS IN OTHER TICK TISSUES

Proteases. Cubilin-related serine proteinase from *H. longicornis* was targeted for RNAi resulting in a reduction in tick weight, presumably through a reduced capacity for the tick to convert blood proteins (Miyoshi *et al.*, 2004). The RNAi-targeted knockdown of several midgut-specific aspartyl endopeptidases (legumains) in *H. longicornis* resulted in an increase in tick mortality, damage to midgut tissues, lower body weight, delayed egg laying, reduced egg viability and reduced oviposition. These data suggest that legumains are involved in digesting the blood meal and remodelling cellular midgut tissues (Alim *et al.*, 2009). A cysteine peptidase Cathepsin L identified in *I. ricinus* was targeted for RNAi resulting in reduced tick engorgement. Cathepsin may be a part of a gut-associated multipeptidase complex that is responsible for digesting the blood meal (Franta *et al.*, 2011). Finally, RNAi-targeted gene knockdown of Ubiquitin/Ubiquitin-63E in *R. annulatus* and *R. microplus* resulted in higher mortality and reduced/nil oviposition (Almazan *et al.*, 2010; Lew-Tabor *et al.*, 2011).

6.6 VACCINE TARGETS

Significant efforts in the tick research community have been directed towards identifying potential targets for an anti-tick vaccine. In some cases, RNAi has been used to screen for novel physiological targets to impair tick feeding/pathogen transmission to the host; these targets could potentially fall into a wide variety of cellular activities. Among the most studied vaccine targets is subolesin, which has been studied in *A. americanum*, *D. marginatus*, *D. variabilis*, *I. scapularis*, *R. annulatus*, *R. microplus* and *Rhipicephalus sanguineus*. Ticks injected with dsRNA against subolesin showed reductions in survival, engorged tick weight, oviposition and egg fertility (Almazan *et al.*, 2010; de la Fuente *et al.*, 2005, 2006a,b,c; Kocan *et al.*, 2009; Nijhof *et al.*, 2007). A number of other vaccine targets have been identified included interphase cytoplasm foci protein 45 (2G7), Putative threonyl-tRNA synthetase (2C9), 60S ribosomal protein (2D10), 60S ribosomal protein (2B7), Hl86, ribosomal protein P0, ReeATAQ, Ree86, Bm86/Rs86, Bm91, Insulin-like growth factor binding proteins, organic anion transporter polypeptide, follistatin-related protein, Elongation factor 1-alpha, Glutathione *S*-transferase, Selenoprotein W and Glycogen synthase kinase-3 (Almazan *et al.*, 2010; de la Fuente *et al.*, 2006a, 2011; Fabres *et al.*, 2010; Gong *et al.*, 2008; Liao *et al.*, 2007; Mulenga and Khumthong, 2010a,b; Nijhof *et al.*, 2007, 2010; Zhou *et al.*, 2006).

6.7 OTHER SELECTED STUDIES

Macrophage migration inhibitory factor (MIF) was identified in *A. americanum* and has been proven to inhibit macrophage movement to the same extent as human MIF. This protein is most highly expressed in the midgut in response to tick feeding, though no physiological phenotype was apparent upon RNAi (Bowen *et al.*, 2010).

Ticks express a variety of defensins as a component of the innate immune system, including Varisin, which has been examined in *D. variabilis*. In ticks injected with varisin dsRNA, the antimicrobial activity of the hemolymph plasma was reduced two- to fourfold compared to control ticks, indicating varisin's role in the tick immune response (Hynes *et al.*, 2008).

Vitellogenin receptors are a subfamily of low-density lipoprotein receptors that mediate the uptake of the yolk protein vitellogenin from the hemolymph. In *D. variabilis*, vitellogenin receptors are absent in virgin females but abundant in the ovaries of mated females following drop off. RNAi targeting these receptors resulted in decreased oviposition and brown eggs, probably due to insufficient vitellogenin in the eggs (Mitchell *et al.*, 2007). In *H. longicornis*, eggs did not fully mature and DNA of *Babesia* was not detected in the eggs of dsRNA-VgR injected ticks, indicating that VgR may also play a role in transovarial transmission of this parasite (Boldbaatar *et al.*, 2008). Three vitellogenins were identified in *H. longicornis*; gene knockdowns resulted in lower tick engorgement weight, higher mortality and lower egg weight (Boldbaatar *et al.*, 2008).

A class B scavenger receptor (SRB) from *H. longicornis* was present in salivary gland acini, undifferentiated midgut tissues and ovary oocytes. Recombinant expression of the SRB was recognized by mouse-SRB antisera. Gene knockdown resulted in a reduction in tick engorgement weight, indicating that SRB may play a role in metabolic default and processes including blood feeding, ovipositon and cuticle formation (Aung *et al.*, 2011).

Lysine-ketoglutarate reductase/saccharopine dehydrogenase (LKR/SDH) catalyses the first two steps of lysine catabolism in plants and mammals. LKR/SDH was identified in a *H. longicornis* cDNA library. The transcript was more abundant in unfed and starved tick tissues than in engorged ticks, suggesting LKR/SDH is important for the starved tick. Gene silencing by RNAi strongly decreased egg viability and tick engorgement weight. These effects may be due to roles that LKR/SDH plays in egg production/reproduction or possibly due to pathway interaction involved in balancing water, carbon and nitrogen levels (Battur *et al.*, 2009).

The formation of disulfide bonds between cysteine residues is essential for proper protein folding and stability. At least three protein disulfide isomerases (PDIs) are present in *H. longicornis*. RNAi of PDI-1 impaired tick blood feeding and decreased oviposition. RNAi of PDI-2 resulted in lower tick viability. RNAi of PDI-3 did not have a significant phenotype, unless a double knockdown assay was performed with PDI-1, which resulted in a synergistic effect and impaired tick viability (Liao *et al.*, 2008).

7 RNAi reagents for a high-throughput RNAi screen?

RNAi has proven to be an extremely potent and versatile tool to specifically reduce expression of targeted genes in blood-feeding ticks (Table 1). Widespread adoption of RNAi using dsRNA as an *in vivo* research tool will accelerate as more effective RNAi delivery tools become commercially available in other systems. RNAi screening technologies *in vivo* have been extensively used in the model organisms *C. elegans* and *D. melanogaster* (Perrimon and Mathey-Prevot, 2007; Vastenhouw *et al.*, 2003). These screens have resulted in the identification of many components of cellular pathways implicated in disease. However, these model organisms lack the complex physiological processes present in the blood-feeding arthropods and particularly in prolonged blood-feeders like ticks. Focused RNAi screens in ticks using gene families are a viable option. Although at this moment such screens are limited to single gene knockdown studies, the further development of new public tick RNAi resources will expand the use of *in vivo* focused RNAi screens. An attractive RNAi screen option exists to use tick cell lines to narrow down targets (Kurtti *et al.*, 2008). RNAi has been widely used to selectively deplete the transcriptional expression in various *in vitro* cell-based systems (Boutros and Ahringer, 2008; Echeverri and Perrimon, 2006; Mohr and Perrimon, 2011; Perrimon and Mathey-Prevot, 2007). A new resource, Tick Cell Biobank (http://tickcells.roslin.ac.uk), serves as a repository for tick cell lines for the global tick and tick-borne disease science community and should serve as an excellent centralized resource to develop new high-throughput RNAi screens.

RNAi has been limited in tick research as a tool to reveal gene function. RNAi has the potential to be used in a focused RNAi screen to assign functions to more than 95% of identified gene sequences. After the genome or transcriptome is sequenced, RNAi can be used to target every gene in the genome and screen for specific phenotypes (Chi *et al.*, 2003). A larger scale approach is needed in which dsRNAs can be generated against all known protein families or open reading frames in the *I. scapularis* genome. Ticks lack a genome-wide RNAi screen to study physiological processes or tick–host and tick–pathogen–host interactions. Genome-wide RNAi screening studies provide major insights into the complexity of biological systems because they facilitate the study of complex interacting pathways. Although RNAi tools are more routinely used for genome scale screening in model organisms (*C. elegans*, *Drosophila*), and a key challenge in vector biology is the development of a genome-wide screen, sharing dsRNA reagents to enable comparison of RNAi reagent activities across tick species, cell lines and assays will facilitate improved RNAi reagents availability. High-throughput RNAi screening (e.g. focused RNAi screen to investigate saliva proteins) projects will facilitate research on ticks and tick-borne diseases as they will form the framework for enhancing our understanding of tick biology and molecular mechanisms.

8 RNAi off-target effects

The use of long dsRNAs to induce RNAi has been accepted within the tick research community as a means to identify function in gene knockdown treatments. However, it is clear that long dsRNAs can produce off-target effects resulting from siRNAs targeting unintended mRNAs. Off-target effects have been relatively well studied in the worm and insect model species and can be reduced by avoiding tri-nucleotide CAN (CA[AGCT]) repeats and by selecting sequences that do not contain regions which can cross-hybridize to other sequences (Perrimon *et al.*, 2010). Several *in silico* databases have subsequently been developed to aid researchers in designing RNAi targets, but these are dependent on full genome sequence information. The prediction of off-target effects is complicated by the fact that as few as 11 complementary contiguous nucleotides and 80% overall complementarity of a siRNA is sufficient to trigger silencing in cultured human cells (Jackson *et al.*, 2003).

These facts led Lew-Tabor *et al.* (2011) to perform an elegant study of off-target effects in *R. microplus*. In experiments using a 594, 196 and 109 bp dsRNA to knockdown ubiquitin-63E, all ticks had a similar phenotype; no ticks laid eggs. However, the authors looked at global gene expression by microarray analysis. Of 144 genes down-regulated in response to injection of the 594 bp dsRNA, 10 showed potential sites for off-target binding (on the basis of a stretch of 16 perfectly complementary nucleotides). Similarly, 13 up-regulated targets (of 136) showed potential sites for off-target binding. The 196 bp dsRNA fragment retained only one predicted off-target match, while the 109 bp dsRNA retained no predicted off-target matches. The qPCR results comparing expression of five of the potential off-target genes in response to all three dsRNA treatments showed that three of them were affected by the 594 bp dsRNA, but not by the 196 or 109 bp dsRNAs, suggesting that they represent genuine off-target effects. These results suggest that in the absence of full genome sequence information, researchers applying RNAi to study tick functional genomics should restrict the size of dsRNA to ~100–200 bp to reduce the possibility of generating off-target effects.

9 Designer RNAi?

The tick sialome revealed a variety of salivary genes and proteins and the most abundant tick salivary proteins are members of multi-gene families (Francischetti *et al.*, 2009). Ticks secrete multiple protein isoforms over several days of blood feeding (5–15 days in different tick species). It may be valuable to target a single gene family by designing siRNA based on conserved gene sequences. A 'designer RNAi' approach may facilitate targeting functional compensation effect of protein isoforms.

10 Conclusions

RNAi has proven to be an extremely potent and versatile functional genomics tool to knockdown the transcriptional expression of target genes. In recent years, this technology has emerged as a powerful tool for research in non-model insects and blood feeding arthropods, both to address questions of basic biology and the need to develop better ways to control or eliminate vector-borne infectious agents. RNAi experiments, together with available sialotranscriptome (*I. scapularis* genome) and proteomics can aid in finding new and novel physiological targets for vector control programs. However, many improvements need to be made before genome-wide RNAi/focussed RNAi in ticks will be efficient. Some important factors influencing the success of dsRNA in non-model organisms like ticks include the following: simple delivery method, the concentration/volume of dsRNA, the nucleotide sequence, knowledge of all protein isoforms, the length of the dsRNA, dsRNA dilution factor, the developmental stage of the tick and duration of tick attachment on the host.

11 Future directions

More research is needed to elucidate the RNAi mechanism across tick species to enhance gene silencing. Functional genomics, either gene-by-gene or through genome-wide high-throughput screens in tick species will be the most obvious use of RNAi as a reverse genetics tool. Genome-wide screens can be carried out in tick-derived cell lines, as a first step in the comprehensive analysis of physiological processes. RNAi screens in tick-derived cell lines should be followed by *in vivo* tick experiments. The application of the RNAi tool to tick species could lead to comparative studies of the function of both genes and gene networks. This could provide significant insight into vector competence, host preference and evolutionary processes.

Acknowledgements

The work was partly supported by USDA NIFA award # 2007-35607-20200, AHA award# 09SDG2280207 and US DOS award# PGA-P21049 to S. K. We thank Lacey Sipsey and Rebecca Browning for their help in compiling Table 1, MS-INBRE core facility supported by NIH-NCRR (P20RR016476), Diana Lovejoy for editing services and reviewers for their very helpful comments.
Competing interests: The authors have declared that no competing interests exist.

References

Alarcon-Chaidez, F. J., Sun, J. and Wikel, S. K. (2007). Transcriptome analysis of the salivary glands of *Dermacentor andersoni* Stiles (Acari: Ixodidae). *Insect Biochem. Mol. Biol.* **37**, 48–71.

Alim, M. A., Tsuji, N., Miyoshi, T., Islam, M. K. Huang, X., *et al.* (2007). Characterization of asparaginyl endopeptidase, legumain induced by blood feeding in the ixodid tick *Haemaphysalis longicornis. Insect Biochem. Mol. Biol.* **37**, 911–922.

Alim, M. A., Tsuji, N., Miyoshi, T., Islam, M. K. Hatta, T., *et al.* (2008a). Developmental stage- and organ-specific expression profiles of asparaginyl endopeptidases/legumains in the ixodid tick *Haemaphysalis longicornis. J. Vet. Med. Sci.* **70**, 1363–1366.

Alim, M. A., Tsuji, N., Miyoshi, T., Islam, M. K. Huang, X., *et al.* (2008b). HlLgm2, a member of asparaginyl endopeptidases/legumains in the midgut of the ixodid tick *Haemaphysalis longicornis*, is involved in blood-meal digestion. *J. Insect Physiol.* **54**, 573–585.

Alim, M. A., Tsuji, N., Miyoshi, T., Islam, M. K., Hatta, T. and Fujisaki, K. (2009). Legumains from the hard tick *Haemaphysalis longicornis* play modulatory roles in blood feeding and gut cellular remodelling and impact embryogenesis. *Int. J. Parasitol.* **39**, 97–107.

Aljamali, M. N., Sauer, J. R. and Essenberg, R. C. (2002). RNA interference: applicability in tick research. *Exp. Appl. Acarol.* **28**, 89–96.

Aljamali, M. N., Bior, A. D., Sauer, J. R. and Essenberg, R. C. (2003). RNA interference in ticks: a study using histamine binding protein dsRNA in the female tick *Amblyomma americanum. Insect Mol. Biol.* **12**, 299–305.

Almazan, C., Lagunes, R., Villar, M., Canales, M., Rosario-Cruz, R., Jongejan, F. and de la Fuente, J. (2010). Identification and characterization of *Rhipicephalus Boophilus microplus* candidate protective antigens for the control of cattle tick infestations. *Parasitol. Res.* **106**, 471–479.

Amdam, G. V., Simoes, Z. L., Guidugli, K. R., Norberg, K. and Omholt, S. W. (2003). Disruption of vitellogenin gene function in adult honeybees by intra-abdominal injection of double-stranded RNA. *BMC Biotechnol.* **3**, 1.

Anisuzzaman, Islam, M. K., Alim, M. A., Miyoshi, T., Hatta, T., Yamaji, K., Matsumoto, Y., Fujisaki, K. and Tsuji, N. (2011). Longistatin, a plasminogen activator, is key to the availability of blood-meals for ixodid ticks. *PLoS Pathog.* **7**, e1001312.

Arolas, J. L., Lorenzo, J., Rovira, A., Castella, J. Aviles, F. X., *et al.* (2005). A carboxypeptidase inhibitor from the tick *Rhipicephalus bursa*: isolation, cDNA cloning, recombinant expression, and characterization. *J. Biol. Chem.* **280**, 3441–3448.

Aung, K. M., Boldbaatar, D., Liao, M., Umemiya-Shirafuji, R., Nakao, S., Matsuoka, T., Tanaka, T. and Fujisaki, K. (2011). Identification and characterization of class B scavenger receptor CD36 from the hard tick, *Haemaphysalis longicornis*. *Parasitol. Res.* **108**, 273–285.

Balashov, Y. S. (1972). Bloodsucking ticks (*Ixodoidea*)-vectors of disease of man and animals (English translation). *Misc. Publ. Entomol. Soc. Am.* **8**, 163–376.

Battur, B., Boldbaatar, D., Umemiya-Shirafuji, R., Liao, M., Battsetseg, B., Taylor, D., Baymbaa, B. and Fujisaki, K. (2009). LKR/SDH plays important roles throughout the tick life cycle including a long starvation period. *PLoS One* **4**, e71.

Belles, X. (2010). Beyond *Drosophila*: RNAi in vivo and functional genomics in insects. *Annu. Rev. Entomol.* **55**, 111–128.

Bergman, D. K., Ramachandra, R. N. and Wikel, S. K. (1998). Characterization of an immunosuppressant protein from *Dermacentor andersoni* (Acari: Ixodidae) salivary glands. *J. Med. Entomol.* **35**, 505–509.

Bergman, D. K., Palmer, M. J., Caimano, M. J., Radolf, J. D. and Wikel, S. K. (2000). Isolation and molecular cloning of a secreted immunosuppressant protein from *Dermacentor andersoni* salivary gland. *J. Parasitol.* **86**, 516–525.

Binnington, K. C. (1978). Sequential changes in salivary structure during attachment and feeding of the cattle tick, *Boophilus microplus*. *Int. J. Parasitol.* **8**, 97–115.

Blandin, S., Moita, L. F., Kocher, T., Wilm, M., Kafatos, F. C. and Levashina, E. A. (2002). Reverse genetics in the mosquito *Anopheles gambiae*; targeted disruption of the Defensin gene. *EMBO Rep.* **3**, 852–856.

Boldbaatar, D., Sikalizyo Sikasunge, C., Battsetseg, B., Xuan, X. and Fujisaki, K. (2006). Molecular cloning and functional characterization of an aspartic protease from the hard tick *Haemaphysalis longicornis*. *Insect Biochem. Mol. Biol.* **36**, 25–36.

Boldbaatar, D., Battsetseg, B., Hatta, T., Miyoshi, T., Tsuji, N., Xuan, X. and Fujisaki, K. (2007). Valosin-containing protein from the hard tick *Haemaphysalis longicornis:* effects of dsRNA-mediated HlVCP gene silencing. *Biochem. Cell Biol.* **85**, 384–394.

Boldbaatar, D., Battsetseg, B., Matsuo, T., Hatta, T., Umemiya-Shirafuji, R., Xuan, X. and Fujisaki, K. (2008). Tick vitellogenin receptor reveals critical role in oocyte development and transovarial transmission of *Babesia* parasite. *Biochem. Cell Biol.* **86**, 331–344.

Boldbaatar, D., Umemiya-Shirafuji, R., Liao, M., Tanaka, T., Xuan, X. and Fujisaki, K. (2010). Multiple vitellogenins from the *Haemaphysalis longicornis* tick are crucial for ovarian development. *J. Insect Physiol.* **56**, 1587–1598.

Boutros, M. and Ahringer, J. (2008). The art and design of genetic screens: RNA interference. *Nat. Rev. Genet.* **9**, 554–566.

Bowen, C. J., Jaworski, D. C., Wasala, N. B. and Coons, L. B. (2010). Macrophage migration inhibitory factor expression and protein localization in *Amblyomma americanum* (Ixodidae). *Exp. Appl. Acarol.* **50**, 343–352.

Bowman, A. S. and Sauer, J. R. (2004). Tick salivary glands: function, physiology and future. *Parasitology* **129**, S67–S81.

Bowman, A. S., Coon, L. B., Needham, G. R. and Sauer, J. R. (1997). Tick saliva: recent advances and implications for vector competence. *Med. Vet. Entomol.* **11**, 277–285.

Bucher, G., Scholten, J. and Klinger, M. (2002). Parental RNAi in *Tribolium* (Coleoptera). *Curr. Biol.* **12**, R85–R86.

Carthew, R. W. (2001). Gene silencing by double-stranded RNA. *Curr. Opin. Cell Biol.* **13**, 244–248.

Chalaire, K. C., Kim, T. K., Garcia-Rodriguez, H. and Mulenga, A. (2011). *Amblyomma americanum* (L.) (Acari: *Ixodidae*) tick salivary gland serine protease inhibitor (serpin) 6 is secreted into tick saliva during tick feeding. *J. Exp. Biol.* **214**, 665–673.

Chi, J. T., Chang, H. Y., Wang, N. N., Chang, D. S., Dunphy, N. and Brown, P. O. (2003). Genomewide view of gene silencing by small interfering RNAs. *Proc. Natl. Acad. Sci. USA* **100**, 6343–6346.

Chmelar, J., Anderson, J. M., Mu, J., Jochim, R. C., Valenzuela, J. G. and Kopecky, J. (2008). Insight into the sialome of the castor bean tick, *Ixodes ricinus*. *BMC Genomics* **9**, 233.

Cogoni, C. and Macino, G. (2000). Post-transcriptional gene silencing across kingdom. *Curr. Opin. Genet. Dev.* **10**, 638–643.

Cupp, E. W. (1991). Biology of ticks. *Vet. Clin. North Am. Small Animal Pract.* **21**, 1. Hoskins, J.D. (Ed.).

Dai, J., Narasimhan, S., Zhang, L., Liu, L., Wang, P. and Fikrig, E. (2010). Tick histamine release factor is critical for *Ixodes scapularis* engorgement and transmission of the lyme disease agent. *PLoS Pathog.* **6**, e1001205.

Daszak, P., Cunningham, A. A. and Hyatt, A. D. (2000). Emerging infectious disease of wildlife-threats to biodiversity and human health. *Science* **287**, 443–449.

de la Fuente, J., Almazán, C., Blouin, E. F., Naranjo, V. and Kocan, K. M. (2005). RNA interference screening in ticks for identification of protective antigens. *Parasitol. Res.* **96**, 137–141.

de la Fuente, J., Almazán, C., Naranjo, V., Blouin, E. F., Meyer, J. M. and Kocan, K. M. (2006a). Autocidal control of ticks by silencing of a single gene by RNA interference. *Biochem. Biophys. Res. Commun.* **344**, 332–338.

de la Fuente, J., Almazán, C., Naranjo, V., Blouin, E. F. and Kocan, K. M. (2006b). Synergistic effect of silencing the expression of tick protective antigens 4D8 and rs86 in *Rhipicephalus sanguineus* by RNA interference. *Parasitol. Res.* **99**, 108–113.

de la Fuente, J., Almazán, C., Blas-Machado, U., Naranjo, V., Mangold, A. J., Blouin, E. F., Gortazar, C. and Kocan, K. M. (2006c). The tick protective antigen, 4D8, is a conserved protein involved in modulation of tick blood ingestion and reproduction. *Vaccine* **24**, 4082–4095.

de la Fuente, J., Kocan, K. M., Almazan, C. and Blouin, E. F. (2007). RNAi interference for the study and genetic manipulation of ticks. *Trends Parasitol.* **23**, 427–433.

de la Fuente, J., Manzano-Roman, R., Naranjo, V., Kocan, K. M., Zivkovic, Z., Blouin, E. F., Canales, M., Almazán, C., Galindo, R. C., Step, D. L. and Villar, M. (2010). Identification of protective antigens by RNA interference for control of the lone star tick, *Amblyomma americanum*. *Vaccine* **28**(7), 1786–1795.

de la Fuente, J., Moreno-Cid, J. A., Canales, M., Villar, M., de la Lastra, J. M., *et al.* (2011). Targeting arthropod subolesin/akirin for the development of a universal vaccine for control of vector infestations and pathogen transmission. *Vet. Parasitol.* **181**, 17–22.

Decrem, Y., Mariller, M., Lahaye, K., Blasioli, V., Beaufays, J., Zouaoui Boudjeltia, K., Vanhaeverbeek, M., Cérutti, M., Brossard, M., Vanhamme, L. and Godfroid, E. (2008). The impact of gene knock-down and vaccination against salivary metalloproteases on blood feeding and egg laying by *Ixodes ricinus*. *Int. J. Parasitol.* **38**, 549–560.

Deruaz, M., Frauenschuch, A., Alessandri, A. L., Dias, J. M., Coelho, F. M., Russo, R. C., Ferreira, B. R., Graham, G. J., Shaw, J. P., Wells, T. N., Teixeira, M. M. Power, C. A., *et al.* (2008). Ticks produce highly selective chemokines binding proteins with anti-inflammatory activity. *J. Exp. Med.* **205**, 2019–2031.

Dong, Y. and Friedrich, M. (2005). Nymphal RNAi: systemic RNAi mediated gene knockdown in juvenile grasshopper. *BMC Biotechnol.* **5**, 25.
Dzitoyeva, S., Dimitrijevic, N. and Maner, H. (2001). Intra-abdominal injection of double-stranded RNA into anesthetized adult *Drosophila* triggers RNA interference in the central nervous system. *Mol. Psychiatry* **6**, 665–670.
Echeverri, C. J. and Perrimon, N. (2006). High-throughput RNAi screening in cultured cells: a user's guide. *Nat. Rev. Genet.* **7**, 373–384.
Espinoza-Gomez, F., Newton-Sanchez, O., Flores-Casares, G., De la Cruz-Ruiz, M., Melnikov, V., Austria-Tejeda, J. and Rojas-Larios, F. (2011). Tick paralysis caused by *Amblyomma maculatum* on the Mexican Pacific Coast. *Vector Borne Zoonotic Dis.* **11**, 945–946.
Fabres, A., De Andrade, C. P., Guizzo, M., Sorgine, M. H., Paiva-Silva, G. O., Masuda, A., Vaz Ida, S. Jr. and Logullo, C. (2010). Effect of GSK-3 activity, enzymatic inhibition and gene silencing by RNAi on tick oviposition and egg hatching. *Parasitology* **137**, 1537–1546.
Fire, A., Xu, S., Montgomery, M. K., Kostas, S. A., Driver, S. E. and Mello, C. C. (1998). Potent and specific genetic interference by double-stranded RNA in *Caenorhabdities elegans*. *Nature* **391**, 806–811.
Francischetti, I. M., Mather, T. N. and Ribeiro, J. M. (2003). Cloning of a salivary gland metalloprotease and characterization of gelatinase and fibrin(ogen)lytic activities in the saliva of the Lyme disease tick vector *Ixodes scapularis*. *Biochem. Biophys. Res. Commun.* **305**, 869–875.
Francischetti, I. M., Mather, T. N. and Ribeiro, J. M. (2004). Penthalaris, a novel recombinant five-Kunitz tissue factor pathway inhibitor (TFPI) from the salivary gland of the tick vector of Lyme disease, *Ixodes scapularis*. *Thromb. Haemost.* **91**, 886–898.
Francischetti, I. M., My Pham, V., Mans, B. J., Andersen, J. F., Mather, T. N., Lane, R. S. and Ribeiro, J. M. (2005a). The transcriptome of the salivary glands of the female western black-legged tick *Ixodes pacificus* (Acari: Ixodidae). *Insect Biochem. Mol. Biol.* **35**, 1142–1161.
Francischetti, I. M., Mather, T. N. and Ribeiro, J. M. (2005b). Tick saliva is a potent inhibitor of endothelial cell proliferation and angiogenesis. *Thromb. Haemost.* **94**, 167–174.
Francischetti, I. M., Sa-Nunes, A., Mans, B. J., Santos, I. M. and Ribeiro, J. M. C. (2009). The role of saliva in tick feeding. *Front. Biosci.* **14**, 2051–2088.
Franta, Z., Sojka, D., Frantova, H., Dvorak, J., Horn, M., Srba, J., Tlacko, P., Mares, M., Schneider, E., Craik, C. S., McKerrow, J. K. Caffrey, C. R., *et al.* (2011). IrCL1—the haemoglobinolytic cathepsin L of the hard tick, *Ixodes ricinus*. *Int. J. Parasitol.* **41**, 1253–1262.
Frauenschuh, A., Power, C. A., Deruaz, M., Ferreira, B. R., Silva, J. S., Teixeira, M. M., Dias, J. M., Martin, T., Wells, T. N. and Proudfoot, A. E. (2007). Molecular cloning and characterization of a highly selective chemokines-binding protein from the tick *Rhipicephalus sanguineus*. *J. Biol. Chem.* **282**, 27250–27358.
Gao, X., Shi, L., Zhou, Y., Cao, J., Zhang, H. and Zhou, J. (2011). Characterization of the anticoagulant protein (Rhipilin-1) from *Rhipicephalus haemaphysaloides* tick. *J. Insect Physiol.* **57**, 339–343.
Garcia, S., Billecocq, A., Crance, J. M., Prins, M., Garin, D. and Bouloy, M. (2006). Viral suppressors of RNA interference impair RNA silencing induced by a Semliki Forest virus replicon in tick cells. *J. Virol.* **79**, 8942–8947.
Geppert, M., Goda, Y., Hammer, R. E., Li, C., Rosahl, T. W., Stevens, C. F. and Sudhof, T. C. (1994). Synaptotagmin I: a major calcium sensor for transmitter release at a central synapse. *Cell* **79**, 717–727.

Gong, H., Liao, M., Zhou, J., Hatta, T., Huang, P., Zhang, G., Kanuka, H., Nishikawa, Y., Xuan, X. and Fujisaki, K. (2008). Gene silencing of ribosomal protein P0 is lethal to the tick *Haemaphysalis longicornis*. *Vet. Parasitol.* **151**, 268–278.

Gong, H., Umemiya, R., Zhou, J., Liao, M., Zhang, H., Jia, H., Nishikawa, Y., Xuan, X. and Fujisaki, K. (2009). Blocking the secretion of saliva by silencing the HlYkt6 gene in the tick *Haemaphysalis longicornis*. *Insect Biochem. Mol. Biol.* **39**, 372–381.

Guerrero, F. D., Moolhuijzen, P., Peterson, D. G., Bidwell, S., Caler, E., Bellgard, M., Nene, V. M. and Djikeng, A. (2010). Reassociation kinetics-based approach for partial genome sequencing of the cattle tick, *Rhipicephalus* (*Boophilus*) *microplus*. *BMC Genomics* **11**, 374.

Hajdusek, O., Sojka, D., Kopacek, P., Buresova, V., Franta, Z., Sauman, I., Winzerling, J. and Grubhoffer, L. (2009). Knockdown of proteins involved in iron metabolism limits tick reproduction and development. *Proc. Natl. Acad. Sci. USA* **106**, 1033–1038.

Harnnoi, T., Sakaguchi, T., Nishikawa, Y., Xuan, X. and Fujisaki, K. (2007). Molecular characterization and comparative study of 6 salivary gland metalloproteases from the hard tick, Haemaphysalis longicornis. *Comp. Biochem. Physiol. B Biochem. Mol. Biol.* **147**, 93–101.

Hatta, T., Umemiya, R., Liao, M., Gong, H., Harnnoi, T., Tanaka, M., Miyoshi, T., Boldbaatar, D., Battsetseg, B., Zhou, J., Xuan, X. Tsuji, N., *et al.* (2007). RNA interference of cytosolic leucine aminopeptidase reduces fecundity in the hard tick, *Haemaphysalis longicornis*. *Parasitol. Res.* **100**, 847–854.

Hechemy, K. E., Samsonoff, W. A., Harris, H. L. and Mckee, M. (1992). Adherence and entry of *Borrelia burgdorferi* in Vero cells. *J. Med. Microbiol.* **36**, 229–238.

Hill, C. A. and Wikel, S. K. (2005). The *Ixodes scapularis* Genome Project: an opportunity for advancing tick research. *Trends Parasitol.* **21**, 151–153.

Horn, M., Nussbaumerova, M., Sanda, M., Kovarova, Z. Srba, J., *et al.* (2009). Hemoglobin digestion in blood-feeding ticks: mapping a multipeptidase pathway by functional proteomics. *Chem. Biol.* **16**, 1053–1063.

Hynes, W. L., Stokes, M. M., Hensley, M., Todd, S. M. and Sonenshine, D. E. (2008). Using RNA interference to determine the role of varisin in the innate immune system of the hard tick *Dermacentor variabilis* (Acari: *Ixodidae*). *Exp. Appl. Acarol.* **46**, 7–15.

Imamura, S., da Silva Vaz Junior, I., Sugino, M., Ohashi, K. and Onuma, M. (2005). A serine protease inhibitor (serpin) from *Haemaphysalis longicornis* as an anti-tick vaccine. *Vaccine* **23**, 1301–1311.

Jackson, A. L., Bartz, S. R., Schelter, J., Kobayashi, S. V., Burchard, J., Mao, M., Li, B., Cavet, G. and Linsley, P. S. (2003). Expression profiling reveals off-target gene regulation by RNAi. *Nat. Biotechnol.* **21**, 635–637.

Jaubert-Possamai, S., Le Trionnaire, G., Bonhomme, J. M., Christophides, G. K., Rispe, C. and Tagu, D. (2007). Gene knockdown by RNAi in the pea aphid Acyrthosiphon pisum. *BMC Biotechnol.* **7**, 63.

Kanost, M. R. (1999). Serine proteinase inhibitors in arthropod immunity. *Dev. Comp. Immunol.* **23**, 291–301.

Karim, S., Essenberg, R. C., Dillwith, J. W., Tucker, J. M., Bowman, A. S. and Sauer, J. R. (2002). Identification of SNARE and cell trafficking regulatory proteins in the salivary glands of the lone star tick, *Amblyomma americanum* (L). *Insect Biochem. Mol. Biol.* **32**, 1711–1721.

Karim, S., Ramakrishnan, V. J., Tucker, J. S., Essenberg, R. C. and Sauer, J. R. (2004a). *Amblyomma americanum* salivary glands: double-stranded RNA-mediated gene silencing of synaptobrevin homologue and inhibition of PGE_2 stimulated protein secretion. *Insect Biochem. Mol. Biol.* **34**, 407–413.

Karim, S., Ramakrishnan, V. J., Tucker, J. S., Essenberg, R. C. and Sauer, J. R. (2004b). *Amblyomma americanum* salivary gland homolog of nSec1 is essential for saliva protein secretion. *Biochem. Biophys. Res. Commun.* **324**, 1256–1263.
Karim, S., Miller, N. J., Valenzuela, J. G., Sauer, J. R. and Mather, T. N. (2005). RNAi-mediated gene silencing to assess the role of synaptobrevin and cystatin in tick blood feeding. *Biochem. Biophys. Res. Commun.* **334**, 1336–1342.
Karim, S., Kenny, B., Troiano, E. and Mather, T. N. (2008). RNAi mediated gene silencing in tick synganglia: a proof of concept study. *BMC Biotechnol.* **8**, 30.
Karim, S., Troiano, E. and Mather, T. N. (2010). Functional genomics tool: gene silencing in nymphs and eggs of *Ixodes scapularis* ticks by electroporated dsRNA. *BMC Biotechnol.* **10**, 1.
Karim, S., Singh, P. and Ribeiro, J. M. C. (2011). A deep insight into the sialotranscriptome of the gulf coast tick *Amblyomma* maculatum. *PLoS One* 6, e28525.
Kaufman, W. R. (1989). Tick-host interaction: a synthesis of current concepts. *Parasitol. Today* **5**, 47–56.
Ketting, R. F. (2011). The many faces of RNAi. *Dev. Cell* **20**, 148–161.
Kocan, K. M., de la Fuente, J., Vlouin, E. F. and Garcia-Garcia, J. C. (2004). *Anaplasma marginale* (Rickettsiales: Anaplasmataceae): recent advances in defining host-pathogen adaptations of a tick-borne rickettsia. *Parasitology* **129**, S285–S300.
Kocan, K. M., Zivkovic, Z., Blouin, E. F., Naranjo, V., Almazán, C., Mitra, R. and de la Fuente, J. (2009). Silencing of genes involved in *Anaplasma marginale*-tick interactions affects the pathogen developmental cycle in *Dermacentor variabilis*. *BMC Dev. Biol.* **9**, 42.
Kopacek, P., Hajdusek, O., Buresova, V. and Daffre, S. (2010). Tick innate immunity. *Adv. Exp. Med. Biol.* **708**, 137–162.
Kotsyfakis, M., Sa-Nunes, A., Francischetti, I. M., Mather, T. N., Andersen, J. F., *et al.* (2006). Antiinflammatory and immunosuppressive activity of sialostatin L, a salivary cystatin from the tick *Ixodes scapularis*. *J. Biol. Chem.* **281**, 26298–26307.
Kotsyfakis, M., Karim, S., Andersen, J. F., Mather, T. N. and Ribeiro, J. M. C. (2007). Selective cysteine protease inhibition contributes to blood-feeding success of the tick *Ixodes scapularis*. *J. Biol. Chem.* **282**, 29256–29263.
Kurscheid, S., Lew-Tabor, A. E., Valle, M. R., Bruyeres, A. G., Doogan, V. J., Munderloh, U. G., Guerrero, F. D., Barrero, R. A. and Bellgard, M. I. (2009). Evidence of a tick RNAi pathway by comparative genomics and reverse genetics screen of targets with known loss-of-function phenotypes in *Drosophila*. *BMC Mol. Biol.* **10**, 26.
Kurtti, J. J., Munderloh, U. G., Hayes, S. F., Kruegar, D. E. and Ahlstrand, G. G. (1994). Ultrastructural analysis of the invasion of tick cells by Lyme disease spirochetes (*Borrelia burgdorferi*) *in vitro*. *Can. J. Zool.* **72**, 977–994.
Kurtti, T. J., Mattila, J. T., Herron, M. J., Felsheim, R. F., Baldridge, G. D., Burkhardt, N. Y., Blazar, B. R., Hackett, P. B., Meyer, J. M. and Munderloh, U. G. (2008). Transgene expression and silencing in a tick cell line: a model system for functional tick genomics. *Insect Biochem. Mol. Biol.* **38**, 963–968.
Lee, Y. S., Nakahara, K., Pham, J. W., Kim, K., He, Z., *et al.* (2004). Distinct roles for Drosophila Dicer-1 and Dicer-2 in the siRNA/miRNA silencing pathways. *Cell* **117**, 69–81.
Lew-Tabor, A. E., Kurscheid, S., Barrero, R., Gondro, C., Moolhuijzen, P. M., Valle, M. R., Morgan, J. A. T., Covacin, C. and Bellgard, M. I. (2011). Gene expression evidence for off-target effects caused by RNA interference mediated gene silencing of Ubiquiting-63E in the cattle tick *Rhipicephalus microplus*. *Int. J. Parasitol.* **41**, 1001–1014.
Li, W. F., Ma, G. X. and Zhou, X. X. (2006). Apidaecin-type peptides: biodiversity, structure–function relationships and mode of action. *Peptides* **27**, 2350–2359.

Liao, M., Zhou, J., Hatta, T., Umemiya, R., Miyoshi, T., Tsuji, N., Xuan, X. and Fujisaki, K. (2007). Molecular characterization of *Rhipicephalus (Boophilus) microplus* Bm86 from *Haemaphysalis longicornis* ticks. *Vet. Parasitol.* **146**, 148–157.

Liao, M., Boldbaatar, D., Gong, H., Huang, P., Umemiya, R., Harnnoi, T., Zhou, J., Tanaka, T., Suzuki, H., Xuan, X. and Fujisaki, K. (2008). Functional analysis of protein disulfide isomerases in blood feeding, viability and oocyte development in *Haemaphysalis longicornis* ticks. *J. Insect Physiol.* **38**, 285–295.

Liu, Q., Rand, T. A., Kalidas, S., Du, F., Kim, H. E., Smith, D. P. and Wang, X. (2003). R2d2, a bridge between the initiation and effector steps of the Drosophila RNAi pathway. *Science* **301**, 1921–1925.

Liu, Q., Carmell, M. A., Rivas, F. V., Marsden, C. G., Thomson, J. M., Song, J. J., Hammond, S. M., Joshua-Tor, L. and Hannon, G. J. (2004). Argonaute2 is the catalytic engine of mammalian RNAi. *Science* **305**, 1437–1441.

Mans, B. J. and Neitz, A. W. (2004). Adaptation of ticks to a blood-feeding environment: evolution from a functional perspective. *Insect Biochem. Mol. Biol.* **34**, 1–17.

Mans, B. J., Louw, A. I. and Neitz, A. W. (2002). Evolution of hematophagy in ticks: common origins for blood coagulation and platelet aggregation inhibitors from soft ticks of the genus *Ornithodoros*. *Mol. Biol. Evol.* **19**, 1695–1705.

Mans, B. J., Andersen, J. F., Francischetti, I. M., Valenzuela, J. G., Schwan, T. G., *et al.* (2008). Comparative sialomics between hard and soft ticks: implications for the evolution of blood-feeding behavior. *Insect Biochem. Mol. Biol.* **38**, 42–58.

May, A. P., Whiteheart, S. W. and Weis, W. I. (2001). Unraveling the mechanism of the vesicle transport ATPase NSF, the N-ethylmaleimide sensitive factor. *J. Biol. Chem.* **276**, 21991–21994.

Meister, G. and Tuschl, T. (2004). Mechanisms of gene silencing by double-stranded RNA. *Nature* **431**, 343–349.

Mello, C. C. and Conte, D. Jr. (2004). Revealing the world of RNA interference. *Nature* **431**, 338–342.

Mitchell, R. D. 3rd, Ross, E., Osgood, C., Sonenshine, D. E., Donohue, K. V., Khalil, S. M., Thompson, D. M. and Michael, Roe R. (2007). Molecular characterization, tissue-specific expression and RNAi knockdown of the first vitellogenin receptor from a tick. *Insect Biochem. Mol. Biol.* **37**, 375–388.

Miyoshi, T., Tsuji, N., Islam, M. K., Kamio, T. and Fujisaki, K. (2004). Gene silencing of a cubilin-related serine proteinase from the hard tick *Haemaphysalis longicornis* by RNA interference. *J. Vet. Med. Sci.* **66**, 1471–1473.

Mohr, S. E. and Perrimon, N. (2011). RNAi screening: new approaches, understandings, and organisms. *Wiley Interdiscip. Rev. RNA* doi:10.1022/wrna.110.

Monteiro, R. Q., Rezaie, A. R., Ribeiro, J. M. and Francischetti, I. M. (2005). Ixolaris: a factor Xa heparin-binding exosite inhibitor. *Biochem. J.* **387**, 871–877.

Monteiro, R. Q., Rezaie, A. R., Bae, J. S., Calvo, E. Andersen, J. F., *et al.* (2008). Ixolaris binding to factor X reveals a precursor state of factor Xa heparin-binding exosite. *Protein Sci.* **17**, 146–153.

Mulenga, A. and Khumthong, R. (2010a). Disrupting the *Amblyomma americanum* (L.) CD147 receptor homolog prevents ticks from feeding to repletion and blocks spontaneous detachment of ticks from their host. *Insect Biochem. Mol. Biol.* **40**, 524–532.

Mulenga, A. and Khumthong, R. (2010b). Silencing of three *Amblyomma americanum* (L.) insulin-like growth factor binding protein-related proteins prevents ticks from feeding to repletion. *J. Exp. Biol.* **213**, 1153–1161.

Mulenga, A., Sugino, M., Nakajim, M., Sugimoto, C. and Onuma, M. (2001). Tick-encoded serine proteinase inhibitors (serpins); potential target antigens for tick vaccine development. *J. Vet. Med. Sci.* **63**, 1063–1069.

Mulenga, A., Khumthong, R., Chalaire, K. C., Strey, O. and Teel, P. (2008). Molecular and biological characterization of the *Amblyomma americanum* organic anion transporter polypeptide. *J. Exp. Biol.* **211**, 3401–3409.

Narasimhan, S., Montgomery, R. R., DePonte, K., Tschudi, C., Marcantonio, N., Anderson, J. F., Sauer, J. R., Cappello, M., Kantor, F. S. and Fikrig, E. (2004). Disruption of *Ixodes scapularis* anticoagulation by using RNA interference. *Proc. Natl. Acad. Sci. USA* **101**, 1141–1146.

Nazareth, R. A., Tomaz, L. S., Ortiz-Costa, S., Atella, G. C., Ribeiro, J. M., *et al.* (2006). Antithrombotic properties of Ixolaris, a potent inhibitor of the extrinsic pathway of the coagulation cascade. *Thromb. Haemost.* **9**, 7–13.

Neelakanta, G., Sultana, H., Fish, D., Anderson, J. F. and Fikrig, E. (2010). *Anaplasma phagocytophilum* induces *Ixodes scapularis* ticks to express an antifreeze glycoprotein gene that enhances their survival in the cold. *J. Clin. Invest.* **120**, 3179–3190.

Nene, V., Lee, D., Quackenbush, J., Skilton, R., Mwaura, S., Gardner, M. J. and Bishop, R. (2002). AvGI, an index of genes transcribed in the salivary glands of the ixodid tick *Amblyomma variegatum*. *Int. J. Parasitol.* **32**, 1447–1456.

Nijhof, A. M., Balk, J. A., Postigo, M., Rhebergen, A. M., Taoufik, A. and Jongejan, F. (2010). Bm86 homologues and novel ATAQ proteins with multiple epidermal growth factor (EFF)-like domains from hard and soft ticks. *Int. J. Parasitol.* **40**, 1587–1597.

Nijhof, A. M., Taoufik, A., de la Fuente, J., Kocan, K. M. de Vries, E., *et al.* (2007). Gene silencing of the tick protective antigens, Bm86, Bm91 and subolesin, in the one-host tick *Boophilus microplus* by RNA interference. *Int. J. Parasitol.* **37**, 653–662.

Oliveira, C. J., Sa-Nunes, A., Francischetti, I. M., Carregaro, V., Anatriello, E., Silva, J. S., Santos, I. K., Ribeiro, J. M. and Ferreira, B. R. (2011). Deconstructing tick saliva: non-protein molecules with potent immunomodulatory properties. *J. Biol. Chem.* **286**, 109600–109609.

Paffenbarger, R. S. Jr. (1951). Tick paralysis: implicating *Amblyomma maculatum*. *New Orleans Med. Surg. J.* **103**, 329–332.

Pal, U., Li, X., Wang, T., Montegomery, R. R., Ramamoorthi, *et al.* (2004). TROSPA, an *Ixodes scapularis* receptor for *Borrelia burgdorferi*. *Cell* **119**, 457–468.

Parola, P. and Raoult, D. (2001). Ticks and tick-borne bacterial diseases in humans: an emerging infectious threat. *Clin. Infect. Dis.* **32**, 897–928.

Pedra, J. H., Narasimhan, S., Deponte, K., Marcantonio, N., Kantor, F. S. and Fikrig, E. (2006). Disruption of the salivary protein 14 in *Ixodes scapularis* nymphs and impact on pathogen acquisition. *Am. J. Trop. Med. Hyg.* **75**, 677–682.

Perrimon, N. and Mathey-Prevot, B. (2007). Matter arising: off-targets and genome-scale RNAi screens in *Drosophila*. *Fly* **1**, 1–5.

Perrimon, N., Ni, J. Q. and Perkins, L. (2010). In vivo RNAi: today and tomorrow. *Cold Spring Harb. Perspect. Biol.* **2**, a003640.

Ramakrishnan, V. G., Aljamali, M. N., Sauer, J. R. and Essenberg, R. C. (2005). Application of RNA interference in tick salivary gland research. *J. Biomol. Tech.* **16**, 297–305.

Rawlings, N. D., Tolle, D. P. and Barrett, A. J. (2004). Evolutionary families of peptidase inhibitors. *Biochem. J.* **378**, 705–716.

Reinhart, B. J., Slack, F. J., Basson, M., Pasquinelli, A. E., Bettinger, J. C., Rougvie, A. E., Horvitz, H. R. and Ruvkun, G. (2000). The 21-nucleotide let-7 RNA regulates developmental timing in *Caenorhabditis elegans*. *Nature* **403**, 901–906.

Ribeiro, J. M. and Francischetti, I. M. (2003). Role of arthropod saliva in blood feeding: sialome and post-sialome perspectives. *Annu. Rev. Entomol.* **48**, 73–88.

Ribeiro, J. M., Alarcon-Chaidez, F., Francischetti, I. M., Mans, B. J., Mather, T. N., Valenzuela, J. G. and Wikel, S. K. (2006). An annotated catalog of salivary gland transcripts from *Ixodes scapularis* ticks. *Insect Biochem. Mol. Biol.* **36**, 111–129.

Ribeiro, J. M., Anderson, J. M., Manoukis, N. C., Meng, Z. and Francischetti, I. M. (2011). A further insight into the sialome of the tropical bont tick, *Amblyomma variegatum*. *BMC Genomics* **12**, 136.

Rothman, J. E. (1994). Mechanisms of intracellular protein transport. *Nature* **372**, 55–63.

Sa-Nunes, A., Bafica, A., Lucas, D. A., Conrads, T. P., Veenstra, T. D., Andersen, J. F., Mather, T. N., Ribeiro, J. M. and Francischetti, I. M. (2007). Prostaglandin E2 is a major inhibitor of dendritic cell maturation and function in *Ixodes scapularis* saliva. *J. Immunol.* **179**, 1497–1505.

Sauer, J. R., McSwain, J. L., Bowman, A. S. and Essenberg, R. C. (1995). Tick salivary gland physiology. *Annu. Rev. Entomol.* **40**, 245–267.

Sauer, J. R., Essenberg, R. C. and Bowman, A. S. (2000). Salivary glands in ixodid ticks: control and mechanism of secretion. *J. Insect Physiol.* **46**, 1069–1078.

Soares, C. A., Lima, C. M., Dolan, M. C., Piesman, J., Beard, C. B. and Zeidner, N. S. (2005). Capillary feeding of specific dsRNA induces silencing of the isac gene in nymphal *Ixodes scapularis* ticks. *Insect Mol. Biol.* **14**, 443–452.

Sojka, D., Francischetti, I. M., Calvo, E. and Kotsyfakis, M. (2007). Cysteine proteases from bloodfeeding arthropod ectoparasites. *Adv. Exp. Med. Biol.* **712**, 177–191.

Sonenshine, D. E. (1991). Biology of Ticks. Vol. 1. Oxford University Press, New York.

Sonenshine, D. E. (1993). Biology of Tick. Vol. 2. Oxford University Press, New York.

Song, J. J., Smith, S. K., Hannon, G. H. and Joshua-Tor, L. (2004). Crystal structure of Argonaute and its implications for RISC slicer activity. *Science* **305**, 1434–1437.

Tabara, H., Yigit, E., Siomi, H. and Mello, C. C. (2002). The dsRNA binding protein RDE-4 interacts with RDE-1, DCR-1, and a DExH-box helicase to direct RNAi in C. elegans. *Cell* **109**, 861–871.

Tomoyasu, Y., Miller, S. C., Tomita, S., Schoppmeier, M. Grossman, D., *et al.* (2008). Exploring systemic RNA interference in insects: a genome-wide survey for RNAi genes in *Tribolium*. *Genome Biol.* **9**, R10.

Valenzuela, J. G., Francischetti, I. M., Pham, V. M., Garfield, M. K., Mather, T. N. and Ribeiro, J. M. (2002). Exploring the sialome of the tick *Ixodes scapularis*. *J. Exp. Biol.* **205**, 2843–2864.

Vastenhouw, N. L., Fischer, S. E., Robert, V. J., Thijssen, K. L., Fraser, A. G., Kamath, R. S., Ahringer, J. and Plasterk, R. H. (2003). A genome-wide screen identifies 27 genes involved in transposon silencing in *C. elegans*. *Curr. Biol.* **13**, 1311–1316.

Wang, M., Guerrero, F. D., Pertea, G. and Nene, V. M. (2007). Global comparative analysis of ESTs from the southern cattle tick, Rhipicephalus (Boophilus) microplus. *BMC Genomics* **12**, 368.

Wheller, W. C. and Hayashi, C. Y. (1998). The phylogeny of the extant chelicerate orders. *Cladistics* **14**, 173–198.

Zhou, J., Liao, M., Hatta, T., Tanaka, M., Xuan, X. and Fujisaki, K. (2006). Identification of a follistatin-related protein from the tick *Haemaphysalis longicornis* and its effect on tick oviposition. *Gene* **372**, 191–198.

Methods for Identifying Small RNA Genes

Fei Li,*,† Fei Ma,‡ Xinglai Ji,§ Shuping Wang,* Yang Sun,* Lun Yu,* Guoqing Li,* and Zhaojun Han*

*Department of Entomology, College of Plant Protection, Nanjing Agricultural University, Nanjing, China

†Key Laboratory of Integrated Management of Crop Diseases and Pests, Nanjing Agricultural University, Ministry of Education, Nanjing, China

‡Laboratory of Comparative Genomics and Bioinformatics & Jiangsu Key Laboratory for Biodiversity and Biotechnology, College of Life Science, Nanjing Normal University, Nanjing, China

§Laboratory for Conservation and Utilization of Bio-Resources & Key Laboratory for Microbial Resources of the Ministry of Education, Yunnan University, Kunming, China

ADVANCES IN INSECT PHYSIOLOGY VOL. 42
ISBN 978-0-12-387680-5
DOI: 10.1016/B978-0-12-387680-5.00005-7

Abstract

As ubiquitous regulators of gene expression, small RNAs have been found in almost all eukaryotic organisms. Identification of small RNA genes remains an important but difficult task. In this review, we summarize strategies for finding small RNA genes using computational approaches, with an emphasis on microRNAs (miRNAs). Methods for preprocessing different types of sequence data, which are necessary to increase the reliability and efficiency, are introduced. Homology searches for conserved miRNAs and strategies for *de novo* prediction of novel miRNAs are described in detail. Because of the high false-positive rates of these approaches, experimental validation of predicted miRNAs is necessary. Northern blots and quantitative real-time PCR (qPCR) analyses are good choices for validating the expression of predicted miRNAs *in vitro* on a small scale, whereas reporter assays are suitable for *in vivo* studies. Because of its higher sensitivity, microarray analysis is efficient for examining the expression levels of predicted miRNAs at a large scale. Algorithms for predicting miRNA targets are also summarized. Methods such as reporter assays, small RNA knockdown, ablation of miRNA biogenesis proteins, and microarrays have been developed to validate target sites either one-by-one or at a high-throughput level. Furthermore, we present a brief introduction to methods for identifying Piwi-interacting RNAs either by experiments or by computational prediction. The advantages and disadvantages of the different strategies and techniques are discussed.

1 Introduction

Non-coding RNA (ncRNA) genes have attracted increasing attentions in the past decade because of their wide existence and important functions in post-transcriptional regulation. Although messenger RNAs (mRNAs) that encode proteins have long been the major players in the molecular biology, two scientific discoveries made the ncRNA to be a rising star. One is that a new type of ncRNA, named microRNA (miRNA), was discovered in a wide range of organisms (Pasquinelli et al., 2000). Another is that tiling array analysis demonstrated that about 85% of the fly genome is transcribed (Manak et al., 2006). These discoveries challenge the traditional understanding that functional genes represent only a small fraction of the genome, suggesting instead the wide existence of ncRNA transcripts.

Based on the lengths of the final products, ncRNAs can be classified into two types: long non-coding RNAs (lncRNAs) and short non-coding RNAs.

lncRNAs such as telomeric repeat-containing RNAs (TERRAs) and large intergenic non-coding RNAs (lincRNAs) have final products with lengths greater than 200 nucleotides (nt) (Azzalin et al., 2007; Blackshaw et al., 2004; Guttman et al., 2009; Li et al., 2011; Wilusz et al., 2009). A database, Non-coding RNA Expression Database (NRED), has been constructed to collect lncRNAs (Dinger et al., 2009). For more details, see the review by Wilusz et al. (2009). The lncRNAs still need be thoroughly investigated and are not within the scope of this review.

Short ncRNAs are a heterogeneous set of RNAs that have lengths less than 40 nt. Among short ncRNAs, small RNA genes such as miRNAs and Piwi-interacting RNAs (piRNAs) have received much attention because they are important post-transcriptional regulators. The term "small RNA genes" specifically refers to endogenous short non-coding genes that have final products less than 40 nt and that function by targeting other RNA molecules such as mRNA or transposons via RNA-silencing mechanisms (Girard et al., 2006; Lagos-Quintana et al., 2001; Lau et al., 2001; Lee and Ambros, 2001). The importance of these types of small RNAs was first recognized in 2001, and small RNAs were named *Science*'s "Break-through of the Year" in 2002 (Science, 2001, 2002, 2003). Other categories of short ncRNAs include small nuclear RNAs (snRNAs), small nucleolar RNAs (snoRNAs), and small cytoplasmic RNAs (scRNAs).

Discovering small RNAs is the first step in studying this type of ncRNA. However, the strategies for identifying small RNA are quite different from those for protein-coding genes. In this review, we first introduce the computational pipelines that have been used for finding small RNA genes in a variety of organisms. The principles of available algorithms are described in order to highlight their advantages and disadvantages. Computational predictions of small RNAs require experimental validation. The experimental methods for detecting small RNAs, target prediction and validation are introduced to provide a systematic overview of the strategies for uncovering the mystery of the small RNAs.

1.1 CLASSIFICATION OF SMALL RNA GENES

To date, at least eight kinds of small RNAs have been described (Table 1). Of these, miRNAs and piRNAs are the most prominent small RNA genes. At present, 18,226 precursor miRNAs (pre-miRNA), expressing 21,643 mature miRNAs, from 168 species, are available in miRbase (version 18). Among them, 2631 miRNAs were found in 24 insect species (Kozomara and Griffiths-Jones, 2011). In the piRNABank database, there are 10,676 piRNA clusters from human, mouse, rat, fruit fly, and zebrafish, of which 569 clusters are from fruit fly (Sai Lakshmi and Agrawal, 2008). There is no central database available for the other six classes of small RNAs. Because they are the best-characterized type of small RNA, we will focus on miRNAs in the following sections. Methods for identifying piRNAs will also be briefly introduced.

TABLE 1
Classification of small RNAs

Classes	Length (nt)	Biogenesis	Biological function	Species	References
MicroRNA (miRNA)	~22	Cleavage of stem-loop precursors by Drosha and Dicer in animals and plants	Development and cell differentiation	Animals, plants, virus	Bartel et al. (2004)
Piwi-interacting RNA (piRNA)	~30	Primarily from transcripts of mobile DNA elements, including active transposons and their wreckage; found in the *Drosophila* genome	Highly expressed during spermatogenesis, mostly in spermatids	Mammals, fruit fly	Grivna et al. (2006a,b)
Repeat-associated small-interfering RNA (rasiRNA)	24–26 (in plants) 24–27 (in fruit flies)	Cleavage by Dicer from long ncRNA transcripts derived from repetitive sequences	Silencing of transposons, repetitive genes, and viruses	Fruit fly, plants, yeast	Chen et al. (2005b)
Trans-acting siRNA	21 (in *Arabidopsis*)	Corresponds to both sense and antisense strands of a non-coding RNA (At2g27400) that apparently is converted to double-stranded RNA	Directs the cleavage of endogenous mRNAs	Plants	Vazquez et al. (2004)

Scan RNA (scnRNA)	Unknown	Unclear, but associated with chromosome rearrangements	Interacts with Twi1p, a PPD protein required for both sequence elimination and small RNA accumulation in *Tetrahymena*; this type of gene is required for DNA elimination and is involved in histone methylation	*Tetrahymena*	Mochizuki et al. (2002, 2004)
Transcription initiation RNAs (tiRNA)	18 (a modal length)	Derived from sequences on the same strand as the TSS and preferentially associated with G+C-rich promoters	Preferentially associated with G+C-rich promoters; also associated with highly expressed transcripts and sites of RNA polymerase II binding	Human, chicken, and *Drosophila*	Taft et al. (2009)
QDE-2-interacting RNA (qiRNA)	~20–21	Originates mostly from the ribosomal DNA locus; induced by DNA damage	Reacts with Argonaute protein QDE-2	Filamentous fungus	Lee et al. (2009)
Small modulatory RNA (smRNA)	~20	Unknown	Neuronal differentiation	Adult hippocampal neural stem cells of mice, human, etc.	Kuwabara et al. (2004)
Tiny non-coding RNA (tncRNA)	~20–21	Cleavage from unidentified precursors by Dicer	Unknown	Nematode	Ambros et al. (2003b)

1.1.1 *MicroRNA*

miRNA genes were first reported and named in 2001 (Lagos-Quintana et al., 2001; Lau et al., 2001; Lee and Ambros, 2001). They are an important class of small endogenous RNA genes with final products in length of ~22 nt. The primary miRNAs (pri-miRNAs) are transcribed from genomic DNA, normally driven by pol II-type promoters. The pri-miRNAs are processed into pre-miRNA by the enzyme Drosha in the nucleus. The pre-miRNAs are transported to the cytoplasm where they are cleaved by the enzyme Dicer into a short double-stranded miRNA duplex (mature miRNA). The pre-miRNAs can be formed into hairpin structures that are the characterized secondary structures of pre-miRNAs (Gregory et al., 2004; Lee et al., 2004). This common feature has been extensively exploited to develop *de novo* prediction algorithms. Some miRNAs are clustered in the genome, whereas others are located in the intergenic or intronic region independently.

Mature miRNAs target the transcripts of protein-coding mRNAs, either to repress translation mechanisms or to trigger the cleavage of target mRNAs. Positions 2–8 of the miRNA (named as seed sequence) have the key roles in target recognition (Brennecke et al., 2005). To differentiate miRNAs from other small RNA genes (particularly endogenous small-interfering RNAs [siRNAs]), five criteria were proposed (Ambros et al., 2003a): (1) A mature miRNA is about 22 nt in length. (2) A miRNA sequence can be precisely matched to the genome sequence of the studied organism. (3) A hairpin structure can be predicted by an RNA secondary structure prediction program (e.g. mfold or RNAfold); the mature miRNA should be located in the stem region of the hairpin. (4) The precursors of miRNAs accumulate when the Dicer enzyme is knocked down. (5) Normally, mature miRNA sequences have strong phylogenetic conservation.

Since 2003, a great deal has been learned about the evolution of miRNAs (Liu et al., 2008; Marco et al., 2010; Nozawa et al., 2010; Tanzer and Stadler, 2004). Comparative analysis indicates that some miRNA genes are highly conserved among different species. For example, let-7 is conserved across most bilaterian phyla (Pasquinelli et al., 2000; Reinhart et al., 2000). Since then, hundreds of conserved miRNAs have been reported in different species (Lai et al., 2003; Wang et al., 2005; Yue et al., 2008). These conserved miRNAs can be easily found by homology searching. However, other miRNA genes are not evolutionarily conserved. Identification of these unconserved miRNAs relies on *de novo* prediction by algorithms developed based on the principle of machine learning. Among 238 miRNA genes in *Drosophila melanogaster*, miRNA family analysis demonstrated that 173 entries have homologues outside of *Drosophila* species, whereas 65 miRNAs are only known from *Drosophila* (from the miRBase). Thus, both homology searching and *de novo* prediction of miRNAs are indispensable for finding miRNAs in a species of interest.

1.1.2 *Piwi-interacting RNA*

PiRNA transcripts arise from uncharacterized precursors and do not require Dicer processing. Members of this class of small RNA genes form RNA–protein complexes through interactions with Piwi proteins (Girard et al., 2006; Grivna et al., 2006a,b). PiRNA genes have been found in several animal species including *D. melanogaster* (Brennecke et al., 2007; Yin and Lin, 2007), *Caenorhabditis elegans* (Batista et al., 2008; Das et al., 2008), and vertebrates (Houwing et al., 2007; Aravin et al., 2008; Lau et al., 2006). In contrast to miRNAs, piRNAs have no hairpin motifs, and their sequences are not conserved across different species. PiRNAs function in the transcriptional silencing of transposons, retrotransposons, and other genetic elements in germ-line cells (Grivna et al., 2006c; Houwing et al., 2007). The mechanisms of piRNA biogenesis are not yet fully understood, though a ''ping-pong'' model has been proposed (Brennecke et al., 2007; Gunawardane et al., 2007). The most distinctive feature of piRNAs is their size, which ranges from 26 to 31 nt in length.

At present, identifying piRNA genes is still a difficult task because they lack common features at both the sequence and structural levels. Two algorithms based on machine-learning methods have been reported, which will be introduced later.

1.1.3 *Other small RNAs*

Other important small RNAs have also been identified and described in the past decade, including endogenous siRNAs, scan RNAs (scnRNAs) (Mochizuki and Gorovsky, 2004; Mochizuki et al., 2002), transcription initiation RNAs (tiRNAs) (Taft et al., 2009), QDE-2-interacting RNAs (qiRNAs) (Lee et al., 2009), small modulatory RNA (smRNAs) (Kuwabara et al., 2004), and tiny non-coding RNA (tncRNAs) (Ambros et al., 2003b).

Endogenous siRNAs can be divided into several subclasses according to their sequence origins, including repeat-associated siRNAs (rasiRNAs) (Chen et al., 2005b; Shpiz et al., 2009) and *trans*-acting siRNAs (Vazquez et al., 2004). scnRNAs interact with Twi1p, a PPD (PAZ and Piwi Domain) protein required for both sequence elimination and small RNA accumulation in *Tetrahymena*, and are involved in histone methylation (Mochizuki and Gorovsky, 2004; Mochizuki et al., 2002). Tiny RNAs (tiRNAs) have been found in *Drosophila* and vertebrates. They are preferentially associated with GC-rich promoters and are derived from sequences between positions -60 and $+120$ on the same strand as the transcriptional start sites (TSS) (Taft et al., 2009). qiRNAs interact with the argonaute protein QDE-2 in response to DNA damage in *Neurospora crassa*. These RNAs originate mostly from the ribosomal DNA locus (Lee et al., 2009). smRNAs play a critical role in mediating neuronal differentiation and are recognized by NRSF/REST (Kuwabara et al., 2004). tncRNAs are not processed from a miRNA-like hairpin precursor but are similar in length to miRNAs

(~20–21 nt). tncRNAs are not conserved. The expression of these small RNA genes is, in some cases, developmentally regulated (Ambros et al., 2003b).

These types of small RNAs were typically detected by high-throughput sequencing of RNA transcripts under special conditions, and little is known about them. Thus, there are no standard methods for detecting these small RNAs.

1.2 CHALLENGES IN IDENTIFYING SMALL RNA GENES

The strategies for identifying small RNAs are quite different from those for protein-coding genes. First, because small RNA genes do not have common features such as codons or open reading frames (ORFs), it is impossible to design degenerate primers based on conserved domains. Second, most computational algorithms for finding protein-coding genes are based on codon usage and ORF features. Therefore, these protein-coding gene-finding algorithms cannot be used to predict small RNAs (Tran et al., 2009).

In the past decade, researchers have worked out efficient pipelines to identify small RNAs. The strategies for identifying various small RNA genes are generally similar. This enables us to review the identification methods together. The main steps for finding small RNA genes are as follows: (1) prepare the sequence data; (2) predict small RNA candidates by homology searching or *de novo* prediction; (3) validate predicted small RNAs; (4) amplify full-length transcripts of the small RNAs (optional); (5) predict the target(s) of the small RNAs (if applicable); and (6) validate the target(s) of the small RNAs (if applicable) (Fig. 1).

2 Predicting miRNA candidates

2.1 PREPARING THE SEQUENCE DATA

The first step for identifying small RNAs is to obtain a collection of sequences that might contain small RNA transcripts. The genome sequences (if available), transcriptome data, and expressed sequence tags (ESTs) can all be used for discovering small RNA genes. At present, sequences obtained from deep sequencing of a small RNA library are the best choice in terms of reliability and sensitivity.

2.1.1 *Genome data*

Genome sequences contain all genetic information and therefore can be exploited to find small RNA genes. They offer the possibility of finding all small RNA genes, but a high rate of false positives needs to be considered. In addition to the NCBI genome database, a few data centres such as FlyBase (Tweedie et al., 2009), SilkDB (Duan et al., 2010), VectorBase (Lawson et al.,

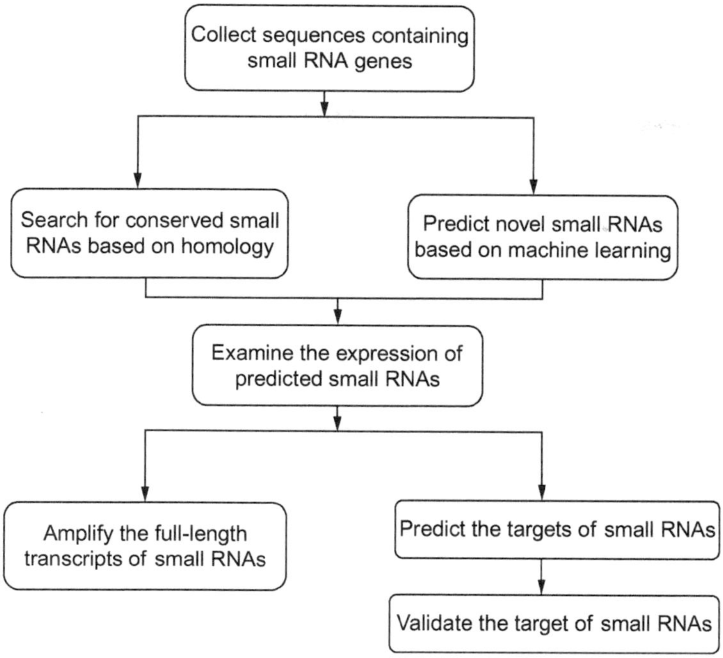

FIG. 1 A simple flowchart for identifying small RNA genes.

2009), and AphidBase (Legeai et al., 2010) provide access to integrated genome resources in insect species (Table 2).

Genome sequences are often well annotated, which is helpful for identifying small RNA candidates. To improve the efficiency and reliability of searches, annotated exons and known non-coding genes are removed and only the unannotated sequences are kept for further small RNA prediction (Fig. 2).

2.1.2 *Transcriptome data and ESTs*

Although transcriptome data and ESTs are used for finding protein-coding genes, increasing evidence suggests that they are also good resources for identifying small RNAs (Jia et al., 2010; Smalheiser, 2003; Xue and Li, 2008; Zhang et al., 2005). Compared with the genome, transcriptome data and ESTs provide expression information that offers high specificity and reliability for detecting small RNAs. The main disadvantage is that only a limited number of highly expressed small RNAs can be found. Publically available transcriptome data are housed at NCBI's GenBank. ESTs can be downloaded from NCBI's database of Expressed Sequence Tags (dbEST).

If the genome sequence of the species of interest is available, the transcriptome data and ESTs need to be mapped to the genome. Using the genome annotations,

TABLE 2
Genome and expressed sequence tag (EST) databases

Species	Database	Link	Description
Silkworm	SilkDB	http://silkworm.genomics.org.cn/	SilkDB is a database that collects gene and genome resource of the silkworm
Drosophila	FlyBase	http://flybase.org/	FlyBase is a comprehensive database that collects genetic and molecular data of *Drosophila*
Mosquito	VectorBase	http://www.vectorbase.org/	VectorBase is a database that collects gene and genome data of mosquitoes
Aphid	AphidBase	http://www.aphidbase.com/aphidbase/ http://w3.rennes.inra.fr/AphidBase/	AphidBase is a database that collects gene and genome data of pea aphid
All	dbEST	http://www.ncbi.nlm.nih.gov/dbEST/	dbEST contains the data of ''expressed sequence tags'', from a number of organisms
All	NCBI genomes database	ftp://ftp.ncbi.nih.gov/genomes/	The NCBI genomes database contains all reported genome data

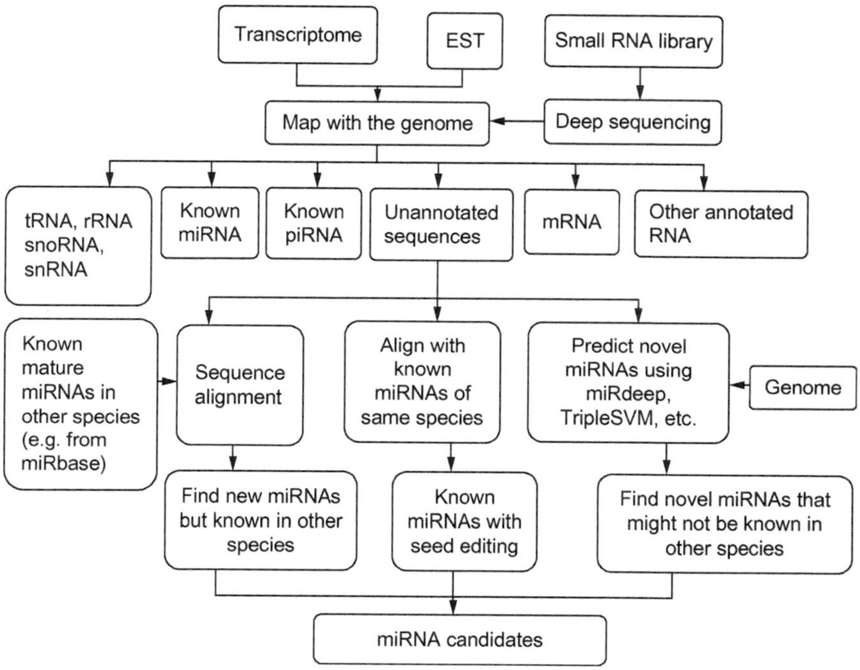

FIG. 2 The pipeline for finding miRNA candidates in organisms with available genome sequences.

known protein-coding genes and other non-coding transcripts such as rRNAs, tRNAs, snRNAs, and snoRNAs should be removed. Only sequences mapped to unannotated regions in the genome should be kept for further analysis (Fig. 2).

If the genome sequence of the species of interest is not available, the transcriptome data and ESTs should be used as the queries to perform BLASTN or BLASTX searches against NCBI's non-redundant (nr) database or other well-organized databases such as SWISS-PROT (Morgulis et al., 2008). The cut-off thresholds can be set as needed. All sequences without any hits to known protein-coding genes or non-coding RNA genes can be used for identifying small RNAs (Fig. 3). The genome sequence of a relative can also be used as the subject sequences in BLAST for annotation. However, we recommend that BLAST against the nr database is generally the better choice.

2.1.3 *Small RNA sequences*

A straightforward approach for obtaining a collection of small RNA sequences is to construct a small RNA library and subject it to deep sequencing (Wei et al., 2009; Zhang et al., 2009). This method is the best and most efficient if cost is not considered. To construct a small RNA library, short RNA fragments should

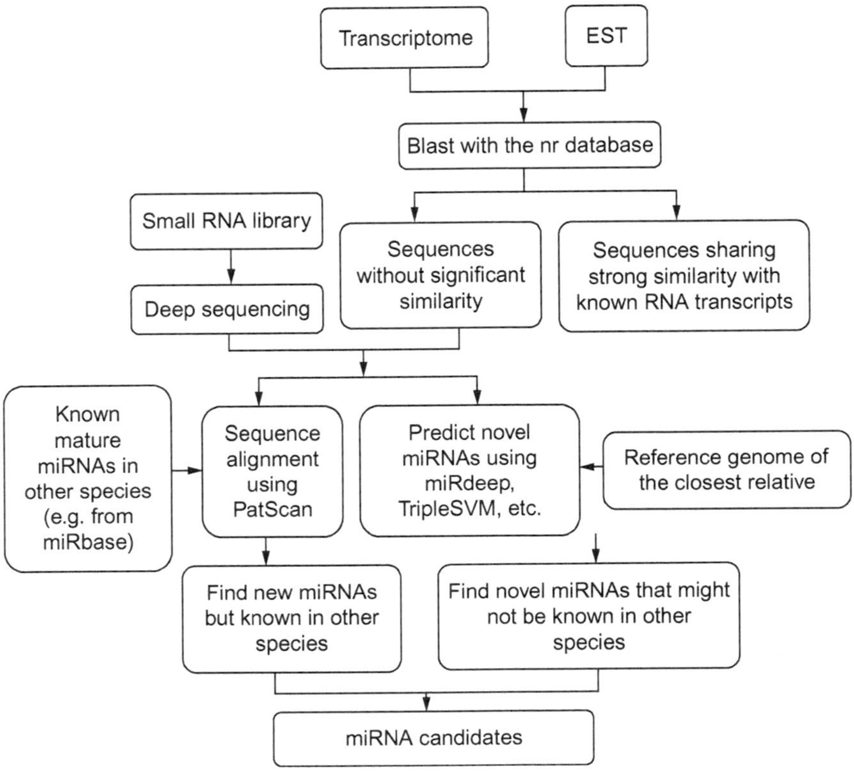

FIG. 3 The pipeline for finding miRNA candidates in organisms without avaliable genome sequences.

be purified from total RNA by extracting low-molecular-weight (LMW) RNA (usually <200 nt) from the samples using commercially available kits (Landgraf et al., 2007) or by recovering short RNA fragments based on the nucleotide length on an acrylamide gel or ion-exchange Q column (Ambros and Lee, 2004; Lau et al., 2006).

Sequences of short RNA fragments obtained by deep sequencing need to be scanned for possible contamination with mRNA sequences because short pieces of degraded mRNA might appear in the small RNA library. PatScan (Dsouza et al., 1997) can be used to match the sequences with the genome (if available) or transcriptome data (if available) to determine whether the short sequences are derived from known protein-coding sequences (Fig. 2). Only short fragments that do not overlap with known protein-coding genes or other non-coding RNA genes are kept. If neither genome sequences nor transcriptome data are available for the species of interest, all short fragment sequences can be directly used for the next step of the analysis, but the reliability and sensitivity need to be considered (Fig. 3).

2.2 HOMOLOGY SEARCHING FOR CONSERVED MIRNAS

Generally, there are two strategies for predicting small RNA candidates: homology searching, which is based on the principle of comparative genomics, and *de novo* prediction, which is based on the principle of machine learning. These two strategies can be used independently, but they are often combined to fully explore the sequences and find more small RNAs. The advantage of homology searching for small RNAs is that it is fast, efficient, and reliable, whereas the disadvantage is that only conserved small RNAs can be found.

The principle of homology searching is that some small RNAs (e.g. mature miRNAs) are evolutionarily conserved across species at the sequence level. To evaluate a query sequence for the presence of a miRNA, this sequence is compared to all known mature miRNAs in the miRBase. If the query sequence shares high similarity with any known miRNA, this query sequence is regarded as a mature miRNA candidate (Fig. 2). If the researchers hope to improve the reliability of the prediction, the seed sequences will be required to perfectly match with the query sequences. The seed sequence extends from the second nucleotide to the seventh or eighth nucleotide of a mature miRNA. It has been demonstrated that the seed sequence plays a key role in miRNA–mRNA interaction (Lewis et al., 2003). However, it should be noticed that the 5′-ends of some miRNAs are not well determined by experiments. In addition, the importance of seed sequence is challenged by further investigation (Brodersen and Voinnet, 2009), and the position of the seed sequence may shift in evolution.

The mature miRNA candidate is mapped to the genome of the same species (if available) or to a reference genome (if the genome of the same species is not available). A long sequence with a length of ~100 nt that contains the query sequence and its flanking sequences is used for secondary structure analysis by RNAfold or mfold. If the long sequence can be folded into a stable stem-loop structure with low free energy (normally <-20kcal/mol), the query sequence is treated as a miRNA candidate. In some cases, the sequences of pre-miRNAs are used for homology searching. If a reference genome is used, the reliability of the analysis decreases. Thus, we do not recommend this strategy for finding miRNAs because of the greater likelihood of false positives.

If there are only a few query sequences, online searches of the miRBase using BLAST are convenient. However, if there are thousands or even millions of query sequences, the speed of searches will be greatly enhanced by constructing a local database of miRNAs and searching for conserved miRNAs using a Perl script. The search parameters can be set based on the evolutionary distance between the species (the species from which the query and subject sequences came). In our work involving searches for conserved miRNAs in insect ESTs, we set up a strict criterion that allows a maximum of three or four edits (one or two mismatches/one deletion/one insertion) (Jia et al., 2010). In insects, many miRNAs have been identified using this strategy (Cao et al., 2008; Huang et al., 2010; Weaver et al., 2007).

Conserved clusters of small RNAs are also important information. Small RNAs, especially miRNAs and piRNAs, are often found in clusters throughout genomes. In some cases, although the sequences of small RNAs are not evolutionarily conserved across species, the pattern of clustering is highly conserved. Therefore, searching for small RNAs near previously known small RNAs and searching for small RNA clusters across the genome can be fruitful (He et al., 2008b).

2.3 *DE NOVO* PREDICTION OF MIRNAS

Although some small RNAs are conserved at the sequence level, there are many small RNAs with low levels of conservation across species. Because homology searching is ineffective in these cases, many algorithms have been developed to predict small RNA candidates *de novo*. The main advantages of *de novo* prediction are that more and novel small RNA candidates can be found, whereas the main disadvantages are low reliability and accuracy.

The principle of this strategy is to exploit common features of secondary structures (e.g. hairpin structures of pre-miRNA), which are learned by the classifiers of different machine-learning methods, such as support vector machine (SVM) and hidden Markov model (HMM). The classifier is normally trained with both positive samples (which have been confirmed to be authentic miRNAs) and negative samples (which are frequently created from exon or shuffled intergenic sequences). After training and testing, the classifier can be used to determine whether a given sequence contains a small RNA gene. Machine-learning methods, such as SVM (Xue et al., 2005), HMM (Agarwal et al., 2010; Oulas et al., 2009), and Naive Bayes (Yousef et al., 2006), have been applied to predict miRNA precursors by considering both structural features and primary sequences.

In general, *de novo* prediction algorithms rely on the evaluation of the secondary structures of pre-miRNAs. As described in Section 2.2, the secondary structure of a given sequence (~100 nt in length) is predicted by RNAfold or mfold. Only the hairpins with low free energies (normally <-20kcal/mol) are kept for evaluation by an SVM or an HMM classifier (Oulas et al., 2009; Zhang et al., 2009).

A sequence of ~100 nt should be obtained before conducting these analyses. For transcriptome data and ESTs whose lengths are normally greater than 100 nt, the ~100-nt sequences are directly extracted from the transcriptome data and ESTs. But for sequences with lengths of only ~40 nt, the methods for *de novo* prediction depend on whether the genome sequence is available or not. If the genome sequence of the focal species is available, the small RNA sequences are first mapped to the genome. A long sequence of ~100 nt that spans the small RNA fragment is extracted and used for evaluation (Fig. 2). If the genome sequence of the species of interest is not available, small RNA sequences should be mapped to a reference genome of a close relative or transcriptome data (if available) to obtain the long sequences (Fig. 3). If a suitable reference genome cannot be found and transcriptome data are also unavailable, small RNA sequences cannot be used for *de novo* prediction of miRNAs with this strategy.

Another strategy to predict miRNAs *de novo* using deep-sequencing data is based on a probabilistic model of miRNA biogenesis. The main principle of this strategy is simple and straightforward but efficient. A hypothesis is proposed that one or more deep-sequencing reads can be mapped to different positions of a sequence if it is a true miRNA precursor. The positions can be divided into three parts—mature miRNA, star sequence, and the loop region—based on the model of process by Dicer. In addition, if this sequence can be folded into a stable hairpin structure with low free energy, it will be treated as a putative pre-miRNA. This strategy is implemented by the software miRDeep and was successfully used in predicting miRNAs (Friedlander et al., 2008). It does not take advantage of machine learning, but it relies on the deep-sequencing data and the genome. If the genome sequence is not available, there is no evidence proving that a reference genome can be used instead. It would yield too many false positives.

At present, it is difficult to summarize the sensitivity and specificity of various algorithms. In most cases, software is chosen based on the interest of the researchers. A list of algorithms is given in Table 3. The reliability of the prediction results mainly depends on experimental validation rather than on the algorithms.

3 Methods for experimentally validating putative miRNAs

The above steps may generate tens or hundreds of small RNA candidates. However, these candidates need to be experimentally confirmed before they are accepted as small RNAs. The principle of validating a small RNA gene is that the expression of a small RNA candidate should be stable under a given situation if it is a true non-coding gene. Otherwise, it might be degraded debris or an incorrectly spliced product of another RNA transcript, even though it meets all of the criteria during the computational screening procedures. In addition, most miRNAs are expressed in a tissue-specific pattern, which makes miRNA detection a difficult task (Liang et al., 2007).

3.1 DETECTING THE EXPRESSION OF SMALL RNAS

The methods for detecting expression of small RNAs are adopted from those for detecting mRNAs with minor modifications. Northern blot and qPCR can be used to study the expression of small RNA candidates *in vitro* on a small scale. *In situ* hybridization (ISH) and reporter assays can be used for detecting the expression of individual small RNA candidates *in vivo*. Microarrays that detect short fragments of small RNA transcripts have also been developed, which have

TABLE 3
Software for predicting miRNAs and piRNAs

Program name	Algorithms	Principle	Species application	Reference	Web site
miRscan	Log-odds score	Comparative genomic	Nematode	Lim et al. (2003)	http://genes.mit.edu/mirscan
miRPara	SVM	*Ab initio* prediction	Animal and plant	Wu et al. (2011)	http://159.226.126.177/mirpara/index.html
mirFinder	SVM	*Ab initio* prediction	Plant	Huang et al. (2007)	http://www.bioinformatics.org/mirfinder
RNAmicro	SVM	*Ab initio* prediction	Animal	Hertel et al. (2006)	http://www.tbi.univie.ac.at/~jana/software/RNAmicro.html
Microprocessor SVM/miRNA SVM	SVM	*Ab initio* prediction	Animal	Helvik et al. (2007)	https://demo1.interagon.com/miRNA
mirEval	SVM	*Ab initio* prediction	Human and mice	Ritchie et al. (2008)	http://tagc.univ-mrs.fr/mireval
TripletSVM	SVM	*Ab initio* prediction	Animal, plant and virus	Xue et al. (2005)	http://bioinfo.au.tsinghua.edu.cn/mirnasvm/
microPred	SVM	*Ab initio* prediction	Human virus	Batuwita et al. (2009)	http://www.comlab.ox.ac.uk/microPred/microPred-server.html
ProMiRII	HMM	*Ab initio* prediction	Animal	Nam et al. (2006)	http://cbit.snu.ac.kr/~ProMiR2
ProMiR	HMM	*Ab initio* prediction	Human	Nam et al. (2006)	http://cbit.snu.ac.kr/~ProMiR2/
miRRim	HMM	*Ab initio* prediction	Animal and plant	Terai et al. (2007)	http://mirrim.ncrna.org/
SSCprofiler	HMM	Comparative genomic	Human	Oulas et al. (2009)	http://www.imbb.forth.gr/SSCprofiler.html

BayesmiRNAfind	NB	*Ab initio* prediction	Animal	Yousef et al. (2006)	https://bioinfo.wistar.upenn.edu/miRNA/miRNA
miPred	RF	*Ab initio* prediction	Human	Jiang et al. (2007)	http://www.bioinf.seu.edu.cn/miRNA/
CID-miRNA	Stochastic context-free grammar	*Ab initio* prediction	Human	Tyagi et al. (2008)	http://mirna.jnu.ac.in/cidmirna/
miRNAkey	–	Comparative genomic	Animal and plant	Ronen et al. (2010)	http://ibis.tau.ac.il/miRNAkey
miRminer	–	Comparative genomic	Animal and plant	Wheeler et al. 2009	http://www.genetics.ucla.edu/labs/horvath/CoexpressionNetwork/Softwares/MiRMiner/
miRseeker	–	Comparative genomic	Fruit fly	Lai et al. (2003)	http://www.fruitfly.org/seq_tools/miRseeker.html
Srnaloop	–	Comparative genomic	Thread worm	Grad et al. (2003)	http://arep.med.harvard.edu/miRNA/pgmlicense.html
MiRAlign	–	Comparative genomic	Animal and plant	Wang et al. (2005)	http://bioinfo.au.tsinghua.edu.cn/miralign/
Vmir	–	*Ab initio* prediction	Virus	Grundhoff et al. (2006)	N/A
MiRDeep	–	*Ab initio* prediction	Animal and plant	Yang et al. (2010)	http://www.mdc-berlin.de/en/research/research_teams/systems_biology_of_gene_regulatory_elements/projects/miRDeep/index.html
mirTools	–	Comparative genomic	Animal and plant	Zhu et al. (2010)	http://centre.bioinformatics.zj.cn/mirtools/
DuplexFinder	–	*Ab initio* prediction	Animal	Zou et al. (2010)	http://dbgroup.cs.tsinghua.edu.cn/zouquan/duplex/

SVM, support vector machine; HMM, Hidden markov model; RF, random forest; NB, Naïve Bayes.

the advantage of being able to examine the expression of hundreds of small RNA genes simultaneously (Fig. 4).

3.1.1 *Northern blot*

Northern blot analysis was used to detect the expression of small RNA genes in the early days, and it is also widely used at present because of its sensitivity and reliability (Lee and Ambros, 2001; McClure et al., 2011). The standard protocols for detecting small RNAs using Northern blots are modified from those for detecting mRNAs. First, the S1 nuclease protection assay was adopted. S1 nuclease from *Aspergillus oryzae* is used to degrade single-stranded oligonucleotides, whereas double-stranded DNA, double-stranded RNA, and DNA–RNA hybrids are resistant to digestion. This apparently increases the sensitivity and specificity of Northern blots (John and Pfeffer, 2011). Second, locked nucleic acid

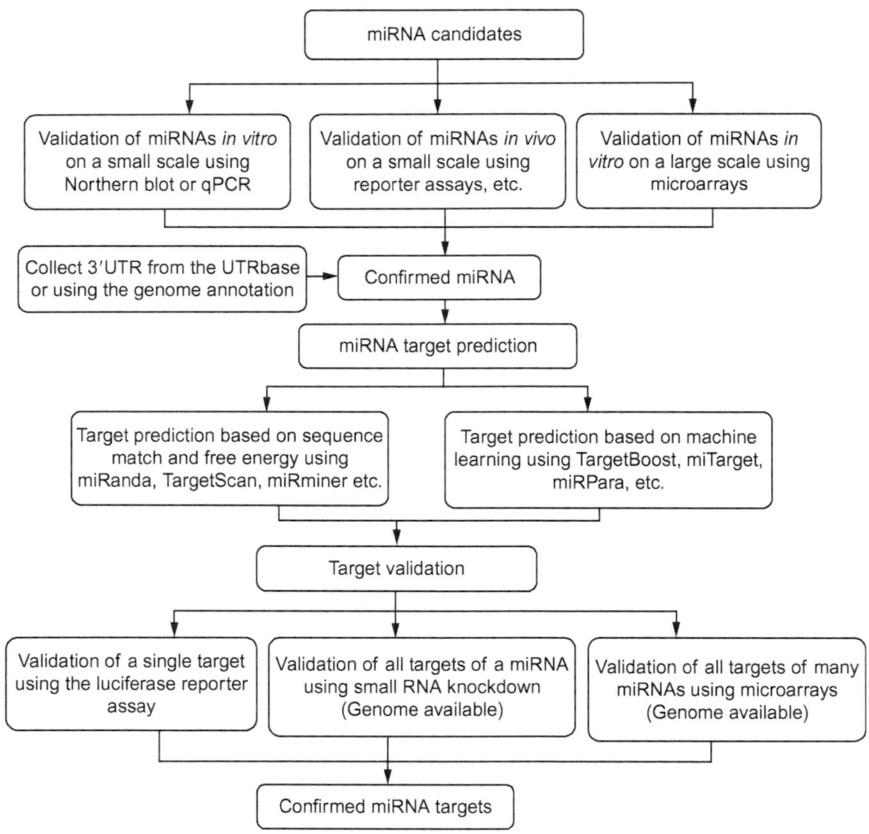

FIG. 4 The pipeline for validating miRNA candidates, as well as for miRNA target prediction and target validation.

(LNA)-modified oligonucleotides were introduced to replace common DNA probes (Valoczi et al., 2004). An LNA nucleotide is a modified RNA nucleotide, with an extra bridge connecting the 2′ oxygen and the 4′ carbon of the sugar group. LNA nucleotides significantly improve hybridization by increasing the thermal stability of duplexes, which is especially important when detecting short RNA or DNA targets. Compared with DNA probes, the sensitivity of Northern blots is increased by at least 10-fold when using LNA probes (Valoczi et al., 2004; Varallyay et al., 2007, 2008). Third, non-radioactive digoxigenin (DIG)-labelled oligonucleotides were substituted for isotope-labeled probes. Compared to ^{32}P-labeled DNA probes, DIG-labelled RNA probes were found to be equally sensitive in detecting miRNAs (Ramkissoon et al., 2006). The combination of DIG-labelling and LNA nucleotides makes the Northern blot a safe, highly specific, and sensitive technique (Kim et al., 2010).

3.1.2 *Quantitative real-time PCR*

Quantitative real-time PCR (qPCR) is an alternative method for highly sensitive, rapid, multiparallel, and cost-effective expression analysis (Udvardi et al., 2008). Some researchers have directly used the protocol for mRNA to detect miRNA expression (Xiong et al., 2009). However, this approach has been challenged for its sensitivity because it detects pri-miRNAs rather than mature miRNAs.

To detect mature miRNAs with qPCR, the standard protocol must be modified. Total RNA or LMW RNA should first be polyadenylated and then reverse-transcribed into cDNA with a poly (T) adapter. The forward primer can be designed based on the sequences of identified small RNAs, and the sequence complementary to the poly (T) adapter is used as the reverse primer. This modified protocol has worked using as little as 100pg of total RNA (Shi and Chiang, 2005). Stem-loop RT-PCR is an alternative and widely used qPCR strategy. A stem-loop RT primer is designed to contain sequences complementary to the 3′-portion of the RNA of interest. The total RNA or LMW RNA is reverse-transcribed to cDNA using the stem-loop RT primer as the anchor primer. Then, the cDNA is quantified using a conventional real-time PCR method. This strategy was invented by Applied Biosystems, which supplies the commercial TaqMan MicroRNA Assay kit. It is highly sensitive and can detect most miRNAs from 25pg of total RNA. This strategy is also applicable for detecting other types of small RNAs (Chen et al., 2005a). Recently, a circularization-based qPCR approach was also proposed. This method relies on the intramolecular ligation of miRNAs and reverse transcription using DNA primers complementary to miRNA circles. After cDNA synthesis, a conventional real-time PCR method using SYBR Green can be used to quantify the RT products (Kumar et al., 2011).

The poly (T) adapter strategy by Shi and Chiang (2005) is cheap and easy to perform in most labs, but its efficiency must be improved. The stem-loop strategy (TaqMan miRNA assay) has high reliability and sensitivity, but it is costly.

3.1.3 In situ *hybridization*

ISH is designed to localize RNA or DNA in fixed tissues by hybridization. An ISH procedure has been developed for detecting miRNAs in mouse embryos and adult organs (Deo et al., 2006). The RNA oligonucleotide probes were conjugated to a fluorescein hapten. The washing conditions were optimized to gain high specificity. The results indicated that ISH detects predominantly mature miRNAs rather than pre-miRNAs or pri-miRNAs. Improvements to standard ISH have been made using similar strategies as described in Section 3.1.1. To increase the sensitivity of hybridization, LNA nucleotides are used to substitute for DNA probes (Debernardi and Dixon-McIver, 2010; Nuovo et al., 2009; Obernosterer et al., 2007; Silahtaroglu et al., 2007). Some researchers have also used DIG-labelled LNA–DNA chimeric probes in ISH procedures because of their sensitivity, specificity, and convenience (Debernardi and Dixon-McIver, 2010; Silahtaroglu, 2010).

3.1.4 *Reporter assay*

Reporter assays are designed to monitor the expression of small RNA genes *in vivo*. A plasmid containing a green fluorescent protein (GFP) reporter/monomeric red fluorescent protein (mRFP) sensor (DFRS) is constructed. The expressions of reporter and sensor are driven by identical constitutive promoters (simian virus 40 [SV40]). The GFP reporter is used to identify the actual expression of the plasmid. The mRFP sensor contains a 3′-untranslated region (UTR) with a tandem cassette that is complementary to the miRNA of interest. Therefore, the mRFP is silenced in the presence of the miRNA of interest. After the plasmid is transferred into living cells, the expression of GFP and mRFP is monitored by fluorescence observation. Using this dual-fluorescent reporter system, the expression of a miRNA can be detected *in vivo* (De Pietri Tonelli et al., 2006).

3.1.5 *Microarrays*

Microarray analysis is a well-known method for studying the expression profiles of protein-coding genes. It has also been adopted to detect the expression of small RNA genes (Krichevsky et al., 2003; Lu et al., 2005; Zhang et al., 2009). The apparent advantage of microarray analysis is that it involves the high-throughput detection of a large number of small RNAs simultaneously, which is impossible to achieve using Northern blot or qPCR.

Linear amplification of target RNA is frequently used to obtain more RNA for microarray analysis. However, a report has demonstrated that although linear amplification of target miRNA increased the sensitivity of the analysis, it also induced a decrease in specificity for closely related probes and increased the variation in dye bias (Saba and Booth, 2006). There are many methods to optimize arrays for detecting small RNAs, such as LNA microarrays

(Fang et al., 2006), bead arrays (Lu et al., 2005), and TaqMan PCR low-density arrays (TLDAs) (Camarillo et al., 2011).

Using different probe design and detection methods, miRNA microarray platforms from different companies were evaluated for their intra-platform repeatability and inter-platform comparability. Although the intra-platform repeatability of different microarray platforms was high, the inter-platform reproducibility was low because of the lack of an adequate normalization method. This means that if a researcher uses different microarray platforms to study miRNA expression, different results might be obtained (Camarillo et al., 2011; Sato et al., 2009; Yauk et al., 2010). A good method for validating microarray results is to use real-time PCR to confirm the expression of the small RNAs of interest.

3.2 AMPLIFYING THE FULL-LENGTH TRANSCRIPTS OF SMALL RNAS

Amplifying the full-length transcripts of small RNAs is an alternative way to validate predicted small RNA candidates. If the full-length transcripts of small RNAs can be successfully amplified, it provides a strong piece of evidence that the predicted small RNAs are true non-coding genes. Rapid amplification of cDNA ends (RACE) was originally developed for amplifying the full-length transcripts of mRNAs. It can be applied to miRNAs because miRNAs are also the products of pol II-type promoters. Many studies have reported that miRNA transcripts can be amplified using the RACE strategy (Bracht et al., 2004; Gabory et al., 2006; Mica et al., 2006; Qian et al., 2011; Xie et al., 2005; Xiong et al., 2009). These results proved that RACE is a good choice to determine the 5′- and 3′-ends of pri-miRNAs. However, the efficiency of this procedure must be improved because of the complex secondary structure and low abundance of pri-miRNA. Recently, we adopted a strategy to increase the abundance of pri-miRNA by stimulating miRNA transcription with $CuSO_4$ and knocking down the Drosha using dsRNA in *Drosophila* S2 cells. RT-PCR proved that the transcripts of pri-miRNA were successfully accumulated, which significantly increase the RACE efficiency (Qian et al., 2011).

4 Methods for miRNA target prediction and validation

4.1 TARGET PREDICTION

The function of small RNA genes is to target other types of RNA molecules with complementary sequences. This complementarity can be used for predicting their targets. Large-scale analysis of miRNA–target duplexes *in vivo* demonstrated that target sites could be divided into broad categories, 5′-dominant sites and 3′-compensatory sites. 5′-dominant sites require perfect or near-perfect base-pairing to the 5′-end of mature miRNA. This class of targets can be further classified into two subtypes, ''canonical sites'' that pair well at both the 5′- and

3′-ends of the miRNA and ''seed sites'' that require little or no base-pair match at the 3′-end. The 3′-compensatory sites have strong compensatory base-pairing at the 3′-end but weak base-pairing at the 5′-end (Brennecke et al., 2005).

Thus, the principles of miRNA target prediction are to consider either 5′-dominant sites or 3′-compensatory sites. The algorithms can be classified into three types according to the region and degree of match considered. The first type of algorithm (e.g. implemented in TargetScan and TargetBoost) considers more of a perfect match between the seed sequence of the mature miRNA and the 3′ UTR of the target mRNA. The second type of algorithm (e.g. implemented in miRanda) considers a perfect match between the sequence at the 5′-end of the mature miRNA (not limited to the seed sequence) and the 3′-UTR of the target mRNA. The third type considers the sequence complementary of the whole mature miRNAs or allows mismatches at the 5′-end. This type of algorithms is mainly developed based on machine-learning methods (e.g. implemented in RNA22). At present, it is difficult to evaluate which kind of algorithm is more reliable and sensitive. These principles are still under challenge by biochemical and molecular biology studies (Brodersen and Voinnet, 2009; Didiano and Hobert, 2006; Easow et al., 2007; Lewis et al., 2005). For reviews on the principles of target prediction, see Brodersen and Voinnet (2009).

Most target prediction algorithms such as miRanda (John et al., 2004), RNAhybrid (Kruger and Rehmsmeier, 2006), TargetScan and TargetScans (Lewis et al., 2005), MicroInspector (Rusinov et al., 2005), PicTar (Grun et al., 2005), and DIANA-microT (Kiriakidou et al., 2004) are based on sequence complementarity and the free energy of the miRNA:mRNA duplex. To increase reliability, other algorithms also consider the sequence conservation of target regions across species (Grun et al., 2005). Machine-learning methods have also been applied to miRNA target prediction. These algorithms mainly rely on the structural features of the miRNA:mRNA duplex, such as GPboost used in TargetBoost (Saetrom et al., 2005), SVM in miTarget (Kim et al., 2006), and Markov models in RNA22 (Miranda et al., 2006).

Evaluations of these algorithms have demonstrated that miRanda can predict a greater number of target mRNAs than DIANA-microT, PicTar, or TargetScanS (Ru et al., 2007). Unfortunately, it is difficult to systematically evaluate the sensitivity and specificity of the prediction algorithms because only a limited number of true targets are known. In terms of target numbers, miRanda provides more candidates for further experimental validation, whereas other algorithms might have higher sensitivity and accuracy.

Another interesting topic in miRNA target prediction is the discovery of abundant miRNA targets in 5′-UTRs and coding regions (CDSs). In animals, miRNAs were thought to target the 3′-UTRs of mRNAs. However, both experimental and computational works have begun to challenge this understanding (Lewis et al., 2005; Miranda et al., 2006; Nakamoto et al., 2005; Stark et al., 2007). In *C. elegans*, a site complementary to the miRNA let-7 was introduced into the 5′-UTR of luciferase in a gene expression system, resulting in the repression of luciferase translation in the presence of let-7 (Lytle et al., 2007).

In *Homo sapiens*, *Mus musculus*, and *D. melanogaster*, abundant miRNA target sites were predicted in 5′-UTRs and CDSs with high degrees of conservation (Zhou et al., 2009). Some targets were confirmed using the reporter assay (Easow et al., 2007). Human miR-10a targets the 5′-UTR of ribosomal protein mRNAs and stimulates their translation (Orom et al., 2008).

4.2 TARGET VALIDATION

Experimental validation of predicted targets is vital for determining the functions of miRNAs. At present, there are several strategies including reporter assays, miRNA knockdown, ablation of miRNA biogenesis proteins, and gene chip assays. Of these methods, destroying miRNA biogenesis affects the expression of all miRNAs rather than a specific miRNA, making it an inappropriate choice for target validation and for miRNA functional analysis. Reporter assays can be used for any species, but miRNA knockdown and gene chips are suitable only for the genome-available species (Fig. 4).

The reporter assay is often used for analysing promoter activity (Alcaraz-Perez et al., 2008). After minor modifications, the reporter assay was applied to miRNA target validation (Du et al., 2009; Fukuda et al., 2005). There have been many successful applications of the reporter assay in validating miRNA targets. For example, by creating an *EGR1* 3′-UTR luciferase reporter construct, miR-191 was proved to target *EGR1* (Saba et al., 2008).

Alternatively, miRNA knockdown is an efficient strategy for studying miRNA functions. Meister and his colleagues proved that antisense oligonucleotides containing 2′-*O*-methyl oligoribonucleotides (OMe), but not 2′-deoxy oligonucleotides, specifically inhibit miRNA (Meister et al., 2004). This approach has been widely applied for silencing miRNA (Chang et al., 2004; Cheng et al., 2005; O'Donnell et al., 2005). A recent study showed that a peptide nucleic acid (PNA) oligomer is also effective in blocking miRNA activity, and it has superior activity compared with standard OMe oligonucleotides (Fabani and Gait, 2008).

Neither reporter assays nor miRNA knockdown studies can be conducted on a high-throughput level. To overcome this limitation, Ribonucleoprotein ImmunoPrecipitation-gene Chip (RIP-Chip) was designed to detect miRNA targetomes. In one example, antibodies against wild-type human Ago2 were used to pull down all gene transcripts connected to the RNA-induced silencing complex (RISC) in untreated Hodgkin lymphoma (HL) cell lines. Then, by using RIP-Chip with anti-miR-17/20/93/106-treated cells, the mRNA targets of miR-17/20/93/106 were identified at a high-throughput level. This method provided a strategy to identify all the targets of a miRNA of interest. However, In the 1189 gene transcripts identified by using anti-miR-17/20/93/106, only 51% had the predicted target sites and only 19% were predicted by TargetScan, indicating that there was no consistency between the results of RIP-Chip and target prediction. This suggests that the high false positives generated during gene chip and immunoprecipitation need to be carefully considered (Tan et al., 2009).

5 Brief introduction to the methods for identifying piRNAs

Strategies for identifying piRNAs are different than those for miRNAs. Homology searching using BLAST or FASTA and motif searching using MEME (Bailey and Elkan, 1994) are not applicable for piRNA prediction because piRNAs are not evolutionarily conserved at the sequence level. Moreover, *de novo* prediction algorithms for miRNAs cannot be used for piRNAs because the uncharacterized precursors of piRNAs lack the hairpin structure motif.

The methods for discovering piRNAs mainly rely on the characteristics of piRNA genes. First, they are longer than miRNAs, with lengths of 25–30 nt. Second, piRNAs are rich in germ cells. Third, they are clustered in the genome. Based on these characteristics, the experimental procedures for cloning piRNAs are modified accordingly. Total RNAs can be extracted from testes, ovaries, or stem cells. The small RNAs used for deep sequencing are then immunoprecipitated using antibodies that recognize the PIWI proteins. Finally, the small RNA sequences obtained by deep sequencing are mapped to the genome. The short fragments that are clustered in the genome and have lengths of ~33 nt are treated as the putative piRNAs. With similar experimental strategies, many piRNAs have been identified in fruit flies (Vagin et al., 2006; Yin and Lin, 2007), mice (Grivna et al., 2006b), rats (Lau et al., 2006), and zebrafish (Houwing et al., 2007).

The field of piRNA computational prediction is still at an early stage, and we are aware of only two computational algorithms to predict piRNAs. By comparing the genomic piRNA clusters with piRNA sequences, a vector with 21×4 components was constructed and used for training with SVM. A weak discriminating sequence motif was found in the piRNAs. Using the sequence classification algorithm, the accuracy of predicting mouse piRNA was about 61% or 72%, depending on the number of piRNA sequences used for training and testing. The limitation of this algorithm is that it can only be used in genome-available organisms (Betel et al., 2007). Recently, a piRNA prediction algorithm, PiRNA-predictor, was reported. This algorithm uses a *k*-mer scheme, relying on the training sets from non-piRNA and piRNA sequences of five model species. PiRNApredictor does not rely on the genome sequences and has a precision greater than 90% and a sensitivity greater than 60% (Zhang et al., 2011). To the best of our knowledge, it is the first *de novo* piRNA prediction software.

6 Software

Bioinformatics programs are indispensable for identifying small RNA genes. To date, a great number of softwares and tools have been developed to identify small RNAs. Most were initially developed for *H. sapiens* or for model organisms such as *M. musculus*, *Rattus norvegicus*, *D. melanogaster*, or *C. elegans*. However, all of them are applicable to insect species because they are all based on of common features of small RNAs that are shared between insects and other animal species. Here we briefly introduce a few commonly used tools (Table 3).

6.1 MIRSCAN

miRScan is an online computational program for identifying miRNA genes based on characteristic features of the conserved stem loop in miRNA precursors. Given a training set of known miRNAs in *C. elegans*, miRscan predicts miRNA precursors by assigning a log-likelihood score to each input sequence based on several features of the hairpin sequences. Some of the features have been found to be more important than others in identifying miRNAs (Lim et al., 2003).

6.2 MIRMINER

miRMiner is designed to mine the reads sequenced from small RNA libraries based on homology searching. The program search for known miRNAs in new taxa based on three criteria, similar length, seed sequence identity, and non-seed sequence similarity. In addition, the program also finds the conserved reads across multiple taxa and treats these as the novel miRNA families (Sperling et al., 2009; Wheeler et al., 2009).

6.3 MIRPARA

miRPara is a program that identifies miRNAs coding region in genome based on SVM. The SVM classifiers were trained with 77 parameters related to the physical properties of the pre-miRNA and its miRNAs. Three SVM classifiers were obtained by classifying all miRNAs into animal, plant, and overall categories. The accuracy of the miRPara software is up to 80%. This software is applicable for finding novel miRNAs as it is based on machine-learning method (Wu et al., 2011).

6.4 MIRALIGN

miRAlign is an online tool for identifying homologous miRNA genes based on both sequence and structural alignment between the query sequence and the experimentally identified miRNAs. miRAlign introduces structural alignment and allows loose conservation and thus can find distant homologous miRNAs. miRAlign can achieve higher sensitivity than simple BLAST searches (Wang et al., 2005).

6.5 TRIPLETSVM

TripletSVM distinguishes the hairpins of real miRNA precursors from pseudo-hairpins by analysing the triplet elements of the queries using an SVM classifier, which was previously trained using a set of real miRNA precursors and a set of pseudo-miRNA hairpins. The TripletSVM classifier built on human data can achieve up to 90% accuracy in various species other than human (Xue et al., 2005).

6.6 MIRFINDER

miRFinder is a pre-miRNA prediction tool designed for genome-wide, pairwise sequence comparisons in two related species. miRFinder searches for hairpin candidates genome-wide and then excludes the non-robust structures by analysing 18 parameters using the SVM method. miRFinder establishes correlations between different types of mutations and the secondary structures of pre-miRNAs and can accurately distinguish between miRNA and non-miRNA hairpins with better performance (Huang et al., 2007).

6.7 CID-MIRNA

CID-miRNA was developed for identifying miRNA precursors in a given DNA sequence with a web interface. CID-miRNA uses experimentally verified human miRNAs as the training data to develop stochastic context-free grammar model for precursor structures. CID-miRNA generates output that lists all of the potential miRNA precursors (Tyagi et al., 2008).

6.8 MIRDEEP

miRDeep is a package developed to identify miRNAs from deep-sequencing data. miRDeep evaluates the secondary structure of the miRNA precursors and gives the scores using a probabilistic model of miRNA biogenesis. miRDeep can discover both previously known and novel miRNAs (Friedlander et al., 2008).

6.9 MIRANALYZER

miRanalyzer is a web server tool for small RNA identification by analysing deep-sequencing data. miRanalyzer identifies unique reads and their copy numbers and detects previously known miRNAs that have been annotated in miRBase and other libraries. A highly accurate machine-learning algorithm based on the random forest classifier has been implemented to predict novel miRNAs (Hackenberg et al., 2009).

6.10 MIREVAL

mireVal is a comprehensive web tool for simple miRNA prediction in genome sequences. mireVal can conduct a thorough analysis of an input sequence up to 10,000 nt long based on four criteria without user-defined cut-off scores, including secondary structure analysis, conservation analysis, cluster analysis, and miRBase BLAST. mireVal provides a user-friendly web interface for researchers without bioinformatics skills (Ritchie et al., 2008).

6.11 MIRTOOLS

mirTools is a comprehensive web server for discovering miRNAs by analysing high-throughput small RNA-sequencing data. The integration of multiple computational approaches in mirTools allows users to characterize deep-sequencing data by raw data preprocessing, genome location, annotation including known miRNA prediction, novel miRNA prediction, and differential expression analysis (Zhu et al., 2010).

6.12 PIRNAPREDICTOR

piRNApredictor was developed to identify piRNA sequences using a *k*-mer scheme. The training sets are based on the non-piRNA and piRNA sequences from five model species. piRNApredictor may achieve high precision and acceptable sensitivity. piRNApredictor can also be used for identifying piRNAs of non-model organisms without complete genome sequences (Zhang et al., 2011).

7 Databases

The small RNA databases provide not only small RNA sequences and related information but also a few tools for data processing and analysis. In addition to the comprehensive databases, a few specialized databases, such as species-specific small RNA databases, have been developed (Table 4). Here we briefly introduce a few of the commonly used small RNA databases.

7.1 MIRBASE

miRBase is the most useful online database resource for miRNA data, containing miRNA nomenclature information, sequence data, annotations, and target predictions. miRBase provides extra information for facilitating miRNA studies, including genomic location and cluster information. miRBase also provides a web interface for browsing and searching miRNA data (Griffiths-Jones, 2010).

7.2 DEEPBASE

deepBase was developed to facilitate the comprehensive annotation and discovery of small and long ncRNAs from high-throughput next-generation sequencing data. deepBase provides an integrative, interactive, and versatile graphical web interface for analysis and visualization of various sequencing data and

TABLE 4
The main databases of small RNAs and their target genes

Database	Available Web site	Description
ASRP	http://asrp.cgrb.oregonstate.edu	The *Arabidopsis* Small RNA Project Database
miRBase	http://microrna.sanger.ac.uk/	MicroRNA sequences, targets, and gene nomenclature
miRNAMap	http://mirnamap.mbc.nctu.edu.tw/	Genomic maps of microRNA genes and their target genes in mammalian genomes
TarBase	http://www.diana.pcbi.upenn.edu/tarbase (original) http://microrna.gr/tarbase (new)	A comprehensive database of experimentally supported animal microRNA targets
Argonaute	http://www.25ma.uni-heidelberg.de/apps/zmf/argonaute/interface	A database for gene regulation by mammalian microRNAs
miRNApath	http://lgmb.fmrp.usp.br/mirnapath	A database of miRNAs, target genes, and metabolic pathways
miRGen	http://www.diana.pcbi.upenn.edu/miRGen	A database for the study of animal microRNA genomic organization and function
miRGator	http://genome.ewha.ac.kr/miRGator/	An integrated system for functional annotation of microRNAs
miRDB	http://mirdb.org	A microRNA target prediction and functional annotation database with a wiki interface
miR2Disease	http://www.miR2Disease.org	A manually curated database for microRNA deregulation in human disease
UCbase and miRfunc	http://microrna.osu.edu/.UCbase4	A database of ultraconserved sequences and microRNA function
miRecords	http://miRecords.umn.edu/miRecords	An integrated resource for microRNA–target interactions
miROrtho	http://cegg.unige.ch/mirortho	A computational survey of microRNA genes
miRo	http://ferrolab.dmi.unict.it/miro	A miRNA knowledgebase
microPC	http://www.biotec.or.th/isl/micropc	A comprehensive resource for predicting and comparing plant microRNAs

miSolRNA	http://www.misolrna.org	A tomato microRNA relational database
TransmiR	http://cmbi.bjmu.edu.cn/transmir	A transcription factor–microRNA regulation database
PMRD	http://bioinformatics.cau.edu.cn/PMRD	A plant microRNA database
Patrocles	http://www.patrocles.org/	A database of polymorphic miRNA-mediated gene regulation in vertebrates
HOCTAR	http://hoctar.tigem.it/	A unique resource for microRNA target prediction
IntmiR	http://rgcb.res.in/intmir	A complete catalogue of intronic miRNAs in human and mouse
miRTarBase	http://miRTarBase.mbc.nctu.edu.tw/	A database that curates experimentally validated microRNA–target interactions
mESAdb	http://konulab.fen.bilkent.edu.tr/mirna/	MicroRNA expression and sequence analysis database
starBase	http://starbase.sysu.edu.cn/	A database for exploring microRNA–mRNA interaction maps from Argonaute CLIP-Seq and Degradome-Seq data
piRNABank	http://pirnabank.ibab.ac.in/	A web resource on classified and clustered Piwi-interacting RNAs
Rfam	http://www.sanger.ac.uk/Software/Rfam/; http://rfam.wustl.edu/	Annotating non-coding RNAs in complete genomes
NONCODE	http://noncode.bioinfo.org.cn	An integrated knowledge database of non-coding RNAs
RNAdb	http://research.imb.uq.edu.au/RNAdb	A comprehensive mammalian non-coding RNA database
Ebbie	http://bioinformatics.org/ebbie/	Automated analysis and storage of small RNA cloning data using a dynamic web server
NPInter	http://bioinfo.ibp.ac.cn/NPInter	The non-coding RNA and protein-related biomacromolecule interaction database
DSAP	http://dsap.cgu.edu.tw	Deep-sequencing small RNA analysis pipeline
ncRNAimprint	http://rnaqueen.sysu.edu.cn/ncRNAimprint	A comprehensive database of mammalian imprinted non-coding RNAs

facilitates transcriptome research and discovery of novel ncRNAs. The latest release of deepBase contains deep-sequencing data from 185 small RNA libraries from diverse tissues and cell lines of seven organisms, including *D. melanogaster* (Yang et al., 2010).

7.3 PIRNABANK

piRNABank provides a web resource for classified and clustered piRNAs, which collects empirically known piRNA sequences and related information reported in human, mouse, and rat. This database provides a user-friendly web interface and supports comprehensive organism- and chromosome-wise searches. piRNABank also provides a graphical interface for visualizing piRNA or piRNA clusters on a genome-wide map (Sai Lakshmi and Agrawal, 2008). To date, this is one of the most comprehensive databases for piRNAs. Although piRNAs have been found in several insects other than *Drosophila* (Zhang et al., 2011), only 569 piRNAs clusters from fruit fly were collected in this database.

7.4 NONCODE

NONCODE is an integrated knowledge database containing almost all types of ncRNAs that are filtered automatically from the literature or GenBank and then curated manually. All ncRNAs deposited in NONCODE have been confirmed by consulting the literature, and most of them are based on experiments. NONCODE introduces a "process function class" for classifying ncRNAs based on cellular process and function. NONCODE provides a user-friendly web interface for visualization and convenient searching. The latest release (v3.0) contains 363,450 public sequences from 1238 organisms. These ncRNA sequences include 133 traditional classes distributed in 26 cellular processes (He et al., 2008a).

8 Perspectives

During the past decade, rapid progress has been made in identifying, characterizing, and using small RNA genes, with emphasis on RNAi pathway-associated small RNAs (e.g. miRNA, piRNA, and siRNA). But identifying small RNA genes still remains one of the most challenging tasks. In contrast to cloning protein-coding mRNAs, small RNA identification relies on a combination of experimental work and bioinformatics analysis. In 2001, when miRNA genes started to be named, only tens of miRNA genes were reported in model organisms. Now, more than 16,000 miRNA genes from more than 150 species have been documented in miRBase. This is due to rapid development in both sequencing technologies and bioinformatics research. In the field of small RNA identification, we think that the following aspects are worthy of more attentions.

First, most bioinformatics algorithms contain tens of computational steps, and the cut-off thresholds or criteria in each step are usually adjustable. Thus, care must be taken in selecting these values. Second, although small RNA libraries derived from deep sequencing may contain millions of sequences, known small RNA genes only account for a very small proportion of these. Most researches have focused on specific types of small RNAs such as miRNAs or piRNAs. More effort should be devoted to study other types of small RNA genes. A better understanding of their functions and mechanisms of biogenesis may greatly assist in developing computer algorithms for *de novo* identification of more small RNA genes. Third, most insect genome sequences are not available. Even for those insects whose genomes have been sequenced, the genomes are still not well annotated. This situation makes the identification of small RNAs in insects a difficult task. In some cases, a reference genome from the silkworm or fruit fly must be used, but this reduces the reliability and specificity of the analysis. Entomologists who study insects without genome sequences should consider this factor carefully. Homology searching for conserved small RNA candidates is much more reliable in this case, but limits researchers to the discovery of miRNAs that have already been characterized.

Acknowledgements

The authors wish to thank Dr. Elizabeth L. Jockusch for her patience and kind assistance in improving the chapter. It is impossible for us to complete this review without her guidance. We also wish to thank the authors of the papers cited in this review for their contributions to small RNA field. Due to space limitation, it is impossible to cite all references, software and databases. The works of the authors are jointly supported by the transgenic programme (2009ZX08001-002B), the National Basic Research Program of China (2009CB125902, 2010CB126200), the National Science Foundation of China (31171843, 31160021, and 30871636), and the Project of New Century Excellent Talents in Universities (No. NCET-07-0405), SRF for ROCS, SEM, Yunnan Province Innovative Research Team (2011CI), and Foundation for Key Teacher of Yunnan University.

References

Agarwal, S., Vaz, C., Bhattacharya, A. and Srinivasan, A. (2010). Prediction of novel precursor miRNAs using a context-sensitive hidden Markov model (CSHMM). *BMC Bioinformatics* **11**(Suppl. 1), S29.

Alcaraz-Perez, F., Mulero, V. and Cayuela, M. L. (2008). Application of the dual-luciferase reporter assay to the analysis of promoter activity in Zebrafish embryos. *BMC Biotechnol.* **8**, 81.

Ambros, V. and Lee, R. C. (2004). Identification of microRNAs and other tiny noncoding RNAs by cDNA cloning. *Methods Mol. Biol.* **265**, 131–158.

Ambros, V., Bartel, B., Bartel, D. P., Burge, C. B., Carrington, J. C., Chen, X., Dreyfuss, G., Eddy, S. R., Griffiths-Jones, S., Marshall, M., Matzke, M., Ruvkun, G., *et al.* (2003a). A uniform system for microRNA annotation. *RNA* **9**, 277–279.

Ambros, V., Lee, R. C., Lavanway, A., Williams, P. T. and Jewell, D. (2003b). MicroRNAs and other tiny endogenous RNAs in C. elegans. *Curr. Biol.* **13**(10), 807–818.

Aravin, A. A., Sachidanandam, R., Bourc'his, D., Schaefer, C., Pezic, D., Toth, K. F., Bestor, T. and Hannon, G. J. (2008). A piRNA pathway primed by individual transposons is linked to de novo DNA methylation in mice. *Mol. Cell* **31**, 785–799.

Azzalin, C. M., Reichenbach, P., Khoriauli, L., Giulotto, E. and Lingner, J. (2007). Telomeric repeat containing RNA and RNA surveillance factors at mammalian chromosome ends. *Science* **318**, 798–801.

Bailey, T. L. and Elkan, C. (1994). Fitting a mixture model by expectation maximization to discover motifs in biopolymers. *Proc. Int. Conf. Intell. Syst. Mol. Biol.* **2**, 28–36.

Bartel, D. P. (2004). MicroRNAs: genomics, biogenesis, mechanism, and function. *Cell* **116**(2), 281–297.

Batista, P. J., Ruby, J. G., Claycomb, J. M., Chiang, R., Fahlgren, N., Kasschau, K. D., Chaves, D. A., Gu, W., Vasale, J. J., Duan, S., Conte, D. Jr., Luo, S., *et al.* (2008). PRG-1 and 21U-RNAs interact to form the piRNA complex required for fertility in C. elegans. *Mol. Cell* **31**, 67–78.

Batuwita, R. and Palade, V. (2009). microPred: effective classification of pre-miRNAs for human miRNA gene prediction. *Bioinformatics* **25**(8), 989–995.

Betel, D., Sheridan, R., Marks, D. S. and Sander, C. (2007). Computational analysis of mouse piRNA sequence and biogenesis. *PLoS Comput. Biol.* **3**, e222.

Blackshaw, S., Harpavat, S., Trimarchi, J., Cai, L., Huang, H., Kuo, W. P., Weber, G., Lee, K., Fraioli, R. E., Cho, S. H., Yung, R., Asch, E., *et al.* (2004). Genomic analysis of mouse retinal development. *PLoS Biol.* **2**, E247.

Bracht, J., Hunter, S., Eachus, R., Weeks, P. and Pasquinelli, A. E. (2004). Trans-splicing and polyadenylation of let-7 microRNA primary transcripts. *RNA* **10**, 1586–1594.

Brennecke, J., Stark, A., Russell, R. B. and Cohen, S. M. (2005). Principles of microRNA-target recognition. *PLoS Biol.* **3**, e85.

Brennecke, J., Aravin, A. A., Stark, A., Dus, M., Kellis, M., Sachidanandam, R. and Hannon, G. J. (2007). Discrete small RNA-generating loci as master regulators of transposon activity in Drosophila. *Cell* **128**, 1089–1103.

Brodersen, P. and Voinnet, O. (2009). Revisiting the principles of microRNA target recognition and mode of action. *Nat. Rev. Mol. Cell Biol.* **10**, 141–148.

Camarillo, C., Swerdel, M. and Hart, R. P. (2011). Comparison of microarray and quantitative real-time PCR methods for measuring microRNA levels in MSC cultures. *Methods Mol. Biol.* **698**, 419–429.

Cao, J., Tong, C., Wu, X., Lv, J., Yang, Z. and Jin, Y. (2008). Identification of conserved microRNAs in Bombyx mori (silkworm) and regulation of fibroin L chain production by microRNAs in heterologous system. *Insect Biochem. Mol. Biol.* **38**, 1066–1071.

Chang, J., Nicolas, E., Marks, D., Sander, C., Lerro, A., Buendia, M. A., Xu, C., Mason, W. S., Moloshok, T., Bort, R., Zaret, K. S. and Taylor, J. M. (2004). miR-122, a mammalian liver-specific microRNA, is processed from hcr mRNA and may downregulate the high affinity cationic amino acid transporter CAT-1. *RNA Biol.* **1**, 106–113.

Chen, C., Ridzon, D. A., Broomer, A. J., Zhou, Z., Lee, D. H., Nguyen, J. T., Barbisin, M., Xu, N. L., Mahuvakar, V. R., Andersen, M. R., Lao, K. Q., Livak, K. J., *et al.* (2005a). Real-time quantification of microRNAs by stem-loop RT-PCR. *Nucleic Acids Res.* **33**, e179.

Chen, P. Y., Manninga, H., Slanchev, K., Chien, M., Russo, J. J., Ju, J., Sheridan, R., John, B., Marks, D. S., Gaidatzis, D., Sander, C., Zavolan, M., *et al.* (2005b). The developmental miRNA profiles of zebrafish as determined by small RNA cloning. *Genes Dev.* **19**, 1288–1293.

Cheng, A. M., Byrom, M. W., Shelton, J. and Ford, L. P. (2005). Antisense inhibition of human miRNAs and indications for an involvement of miRNA in cell growth and apoptosis. *Nucleic Acids Res.* **33**, 1290–1297.

Das, P. P., Bagijn, M. P., Goldstein, L. D., Woolford, J. R., Lehrbach, N. J., Sapetschnig, A., Buhecha, H. R., Gilchrist, M. J., Howe, K. L., Stark, R., Matthews, N., Berezikov, E., *et al.* (2008). Piwi and piRNAs act upstream of an endogenous siRNA pathway to suppress Tc3 transposon mobility in the Caenorhabditis elegans germline. *Mol. Cell* **31**, 79–90.

De Pietri Tonelli, D., Calegari, F., Fei, J. F., Nomura, T., Osumi, N., Heisenberg, C. P. and Huttner, W. B. (2006). Single-cell detection of microRNAs in developing vertebrate embryos after acute administration of a dual-fluorescence reporter/sensor plasmid. *Biotechniques* **41**, 727–732.

Debernardi, S. and Dixon-McIver, A. (2010). MicroRNA detection in bone marrow cells by LNA-FISH. *Methods Mol. Biol.* **667**, 33–45.

Deo, M., Yu, J. Y., Chung, K. H., Tippens, M. and Turner, D. L. (2006). Detection of mammalian microRNA expression by in situ hybridization with RNA oligonucleotides. *Dev. Dyn.* **235**, 2538–2548.

Didiano, D. and Hobert, O. (2006). Perfect seed pairing is not a generally reliable predictor for miRNA-target interactions. *Nat. Struct. Mol. Biol.* **13**, 849–851.

Dinger, M. E., Pang, K. C., Mercer, T. R., Crowe, M. L., Grimmond, S. M. and Mattick, J. S. (2009). NRED: a database of long noncoding RNA expression. *Nucleic Acids Res.* **37**, D122–126.

Dsouza, M., Larsen, N. and Overbeek, R. (1997). Searching for patterns in genomic data. *Trends Genet.* **13**, 497–498.

Du, T. T., Fu, Y. F., Dong, M., Wang, L., Fan, H. B., Chen, Y., Jin, Y., Chen, S. J., Chen, Z., Deng, M., Huang, Q. H. and Liu, T. X. (2009). Experimental validation and complexity of miRNA-mRNA target interaction during zebrafish primitive erythropoiesis. *Biochem. Biophys. Res. Commun.* **381**, 688–693.

Duan, J., Li, R., Cheng, D., Fan, W., Zha, X., Cheng, T., Wu, Y., Wang, J., Mita, K., Xiang, Z. and Xia, Q. (2010). SilkDB v2.0: a platform for silkworm (Bombyx mori) genome biology. *Nucleic Acids Res.* **38**, D453–456.

Easow, G., Teleman, A. A. and Cohen, S. M. (2007). Isolation of microRNA targets by miRNP immunopurification. *RNA* **13**, 1198–1204.

Fabani, M. M. and Gait, M. J. (2008). miR-122 targeting with LNA/2′-O-methyl oligonucleotide mixmers, peptide nucleic acids (PNA), and PNA-peptide conjugates. *RNA* **14**, 336–346.

Fang, S., Lee, H. J., Wark, A. W. and Corn, R. M. (2006). Attomole microarray detection of microRNAs by nanoparticle-amplified SPR imaging measurements of surface polyadenylation reactions. *J. Am. Chem. Soc.* **128**, 14044–14046.

Friedlander, M. R., Chen, W., Adamidi, C., Maaskola, J., Einspanier, R., Knespel, S. and Rajewsky, N. (2008). Discovering microRNAs from deep sequencing data using miRDeep. *Nat. Biotechnol.* **26**, 407–415.

Fukuda, Y., Kawasaki, H. and Taira, K. (2005). Exploration of human miRNA target genes in neuronal differentiation. *Nucleic Acids Symp. Ser. (Oxf.)* **49**, 341–342.

Gabory, A., Ripoche, M. A., Yoshimizu, T. and Dandolo, L. (2006). The H19 gene: regulation and function of a non-coding RNA. *Cytogenet. Genome Res.* **113**, 188–193.

Girard, A., Sachidanandam, R., Hannon, G. J. and Carmell, M. A. (2006). A germline-specific class of small RNAs binds mammalian Piwi proteins. *Nature* **442**, 199–202.

Grad, Y., Aach, J., Hayes, G. D., Reinhart, B. J., Church, G. M., Ruvkun, G. and Kim, J. (2003). Computational and experimental identification of *C. elegans* microRNAs. *Mol. Cell.* **11**(5), 1253–1263.

Gregory, R. I., Yan, K. P., Amuthan, G., Chendrimada, T., Doratotaj, B., Cooch, N. and Shiekhattar, R. (2004). The Microprocessor complex mediates the genesis of microRNAs. *Nature* **432**, 235–240.

Griffiths-Jones, S. (2010). miRBase: microRNA sequences and annotation. *Curr. Protoc. Bioinformatics* Chapter 12, Unit 12, 19. 11-10.

Grivna, S. T., Beyret, E., Wang, Z. and Lin, H. (2006a). A novel class of small RNAs in mouse spermatogenic cells. *Genes Dev.* **20**(13), 1709–1714.

Grivna, S. T., Pyhtila, B. and Lin, H. (2006b). MIWI associates with translational machinery and PIWI-interacting RNAs (piRNAs) in regulating spermatogenesis. *Proc. Natl. Acad. Sci. USA* **103**, 13415–13420.

Grun, D., Wang, Y. L., Langenberger, D., Gunsalus, K. C. and Rajewsky, N. (2005). MicroRNA target predictions across seven Drosophila species and comparison to mammalian targets. *PLoS Comput. Biol.* **1**, e13.

Grundhoff, A., Sullivan, C. S. and Ganem, D. (2006). A combined computational and microarray-based approach identifies novel microRNAs encoded by human gammaherpesviruses. *RNA* (5), 733–750.

Gunawardane, L. S., Saito, K., Nishida, K. M., Miyoshi, K., Kawamura, Y., Nagami, T., Siomi, H. and Siomi, M. C. (2007). A slicer-mediated mechanism for repeat-associated siRNA 5′ end formation in Drosophila. *Science* **315**, 1587–1590.

Guttman, M., Amit, I., Garber, M., French, C., Lin, M. F., Feldser, D., Huarte, M., Zuk, O., Carey, B. W., Cassady, J. P., Cabili, M. N., Jaenisch, R., *et al.* (2009). Chromatin signature reveals over a thousand highly conserved large non-coding RNAs in mammals. *Nature* **458**, 223–227.

Hackenberg, M., Sturm, M., Langenberger, D., Falcon-Perez, J. M. and Aransay, A. M. (2009). miRanalyzer: a microRNA detection and analysis tool for next-generation sequencing experiments. *Nucleic Acids Res.* **37**, W68–76.

He, S., Liu, C., Skogerbo, G., Zhao, H., Wang, J., Liu, T., Bai, B., Zhao, Y. and Chen, R. (2008a). NONCODE v2.0: decoding the non-coding. *Nucleic Acids Res.* **36**, D170–D172.

He, T., Li, F., Gu, J., Li, R. and Li, F. (2008b). Computational identification of 99 insect microRNAs using comparative genomics. *Tsinghua Sci. Technol.* **13**, 425–432.

Helvik, S. A., Snøve, O., Jr, and Saetrom, P. (2007). Reliable prediction of Drosha processing sites improves microRNA gene prediction. *Bioinformatics* **23**(2), 142–149.

Hertel, J. and Stadler, P. F. (2006). Hairpins in a Haystack: recognizing microRNA precursors in comparative genomics data. *Bioinformatics* **22**(14), e197–e202.

Houwing, S., Kamminga, L. M., Berezikov, E., Cronembold, D., Girard, A., van den Elst, H., Filippov, D. V., Blaser, H., Raz, E., Moens, C. B., Plasterk, R. H., Hannon, G. J., *et al.* (2007). A role for Piwi and piRNAs in germ cell maintenance and transposon silencing in Zebrafish. *Cell* **129**, 69–82.

Huang, T. H., Fan, B., Rothschild, M. F., Hu, Z. L., Li, K. and Zhao, S. H. (2007). MiRFinder: an improved approach and software implementation for genome-wide fast microRNA precursor scans. *BMC Bioinformatics* **17**(8), 341.

Huang, Y., Zou, Q., Tang, S. M., Wang, L. G. and Shen, X. J. (2010). Computational identification and characteristics of novel microRNAs from the silkworm (Bombyx mori L.). *Mol. Biol. Rep.* **37**, 3171–3176.

Jia, Q., Lin, K., Liang, J., Yu, L. and Li, F. (2010). Discovering conserved insect microRNAs from expressed sequence tags. *J. Insect Physiol.* **56**, 1763–1769.

Jiang, P., *et al.* (2007). MiPred: classification of real and pseudo microRNA precursors using random forest prediction model with combined features. *Nucleic Acids Res.* **34**, W339–W344.

John, M. and Pfeffer, S. (2011). Detection of viral microRNA with S1 nuclease protection assay. *Methods Mol. Biol.* **721**, 173–182.

John, B., Enright, A. J., Aravin, A., Tuschl, T., Sander, C. and Marks, D. S. (2004). Human microRNA targets. *PLoS Biol.* **2**, e363.

Kim, S. K., Nam, J. W., Rhee, J. K., Lee, W. J. and Zhang, B. T. (2006). miTarget: microRNA target gene prediction using a support vector machine. *BMC Bioinformatics* **7**, 411.

Kim, S. W., Li, Z., Moore, P. S., Monaghan, A. P., Chang, Y., Nichols, M. and John, B. (2010). A sensitive non-radioactive northern blot method to detect small RNAs. *Nucleic Acids Res.* **38**, e98.

Kiriakidou, M., Nelson, P. T., Kouranov, A., Fitziev, P., Bouyioukos, C., Mourelatos, Z. and Hatzigeorgiou, A. (2004). A combined computational-experimental approach predicts human microRNA targets. *Genes Dev.* **18**, 1165–1178.

Kozomara, A. and Griffiths-Jones, S. (2011). miRBase: integrating microRNA annotation and deep-sequencing data. *Nucleic Acids Res.* **39**, D152–157.

Krichevsky, A. M., King, K. S., Donahue, C. P., Khrapko, K. and Kosik, K. S. (2003). A microRNA array reveals extensive regulation of microRNAs during brain development. *RNA* **9**, 1274–1281.

Kruger, J. and Rehmsmeier, M. (2006). RNAhybrid: microRNA target prediction easy, fast and flexible. *Nucleic Acids Res.* **34**, W451–W454.

Kumar, P., Johnston, B. H. and Kazakov, S. A. (2011). miR-ID: a novel, circularization-based platform for detection of microRNAs. *RNA* **17**, 365–380.

Kuwabara, T., Hsieh, J., Nakashima, K., Taira, K. and Gage, F. H. (2004). A small modulatory dsRNA specifies the fate of adult neural stem cells. *Cell* **116**(6), 779–793.

Lagos-Quintana, M., Rauhut, R., Lendeckel, W. and Tuschl, T. (2001). Identification of novel genes coding for small expressed RNAs. *Science* **294**, 853–858.

Lai, E. C., Tomancak, P., Williams, R. W. and Rubin, G. M. (2003). Computational identification of Drosophila microRNA genes. *Genome Biol.* **4**(7), R42.

Landgraf, P., Rusu, M., Sheridan, R., Sewer, A., Iovino, N., Aravin, A., Pfeffer, S., Rice, A., Kamphorst, A. O., Landthaler, M., Lin, C., Socci, N. D., *et al.* (2007). A mammalian microRNA expression atlas based on small RNA library sequencing. *Cell* **129**, 1401–1414.

Lau, N. C., Lim, L. P., Weinstein, E. G. and Bartel, D. P. (2001). An abundant class of tiny RNAs with probable regulatory roles in Caenorhabditis elegans. *Science* **294**, 858–862.

Lau, N. C., Seto, A. G., Kim, J., Kuramochi-Miyagawa, S., Nakano, T., Bartel, D. P. and Kingston, R. E. (2006). Characterization of the piRNA complex from rat testes. *Science* **313**, 363–367.

Lawson, D., Arensburger, P., Atkinson, P., Besansky, N. J., Bruggner, R. V., Butler, R., Campbell, K. S., Christophides, G. K., Christley, S., Dialynas, E., Hammond, M., Hill, C. A., *et al.* (2009). VectorBase: a data resource for invertebrate vector genomics. *Nucleic Acids Res.* **37**, D583–D587.

Lee, R. C. and Ambros, V. (2001). An extensive class of small RNAs in Caenorhabditis elegans. *Science* **294**, 862–864.

Lee, Y., Kim, M., Han, J., Yeom, K. H., Lee, S., Baek, S. H. and Kim, V. N. (2004). MicroRNA genes are transcribed by RNA polymerase II. *EMBO J.* **23**, 4051–4060.

Lee, H. C., Chang, S. S., Choudhary, S., Aalto, A. P., Maiti, M., Bamford, D. H. and Liu, Y. (2009). qiRNA is a new type of small interfering RNA induced by DNA damage. *Nature* **459**(7244), 274–277.

Legeai, F., Shigenobu, S., Gauthier, J. P., Colbourne, J., Rispe, C., Collin, O., Richards, S., Wilson, A. C., Murphy, T. and Tagu, D. (2010). AphidBase: a centralized bioinformatic resource for annotation of the pea aphid genome. *Insect Mol. Biol.* **19**(Suppl 2), 5–12.

Lewis, B. P., Shih, I. H., Jones-Rhoades, M. W., Bartel, D. P. and Burge, C. B. (2003). Prediction of mammalian microRNA targets. *Cell* **115**, 787–798.

Lewis, B. P., Burge, C. B. and Bartel, D. P. (2005). Conserved seed pairing, often flanked by adenosines, indicates that thousands of human genes are microRNA targets. *Cell* **120**, 15–20.

Li, D., Wang, Y., Zhang, K., Jiao, Z., Zhu, X., Skogerboe, G., Guo, X., Chinnusamy, V., Bi, L., Huang, Y., Dong, S., Chen, R., *et al.* (2011). Experimental RNomics and genomic comparative analysis reveal a large group of species-specific small non-message RNAs in the silkworm Bombyx mori. *Nucleic Acids Res.* **39**, 3792–3805.

Liang, Y., Ridzon, D., Wong, L. and Chen, C. (2007). Characterization of microRNA expression profiles in normal human tissues. *BMC Genomics* **8**, 166.

Lim, L. P., Lau, N. C., Weinstein, E. G., Abdelhakim, A., Yekta, S., Rhoades, M. W., Burge, C. B. and Bartel, D. P. (2003). The microRNAs of Caenorhabditis elegans. *Genes Dev.* **17**(8), 991–1008.

Liu, N., Okamura, K., Tyler, D. M., Phillips, M. D., Chung, W. J. and Lai, E. C. (2008). The evolution and functional diversification of animal microRNA genes. *Cell Res.* **18**, 985–996.

Lu, J., Getz, G., Miska, E. A., Alvarez-Saavedra, E., Lamb, J., Peck, D., Sweet-Cordero, A., Ebert, B. L., Mak, R. H., Ferrando, A. A., Downing, J. R., Jacks, T., *et al.* (2005). MicroRNA expression profiles classify human cancers. *Nature* **435**, 834–838.

Lytle, J. R., Yario, T. A. and Steitz, J. A. (2007). Target mRNAs are repressed as efficiently by microRNA-binding sites in the 5′ UTR as in the 3′ UTR. *Proc. Natl. Acad. Sci. USA* **104**, 9667–9672.

Manak, J. R., Dike, S., Sementchenko, V., Kapranov, P., Biemar, F., Long, J., Cheng, J., Bell, I., Ghosh, S., Piccolboni, A. and Gingeras, T. R. (2006). Biological function of unannotated transcription during the early development of Drosophila melanogaster. *Nat. Genet.* **38**, 1151–1158.

Marco, A., Hui, J. H., Ronshaugen, M. and Griffiths-Jones, S. (2010). Functional shifts in insect microRNA evolution. *Genome Biol. Evol.* **2**, 686–696.

McClure, L. V., Lin, Y. T. and Sullivan, C. S. (2011). Detection of viral microRNAs by Northern blot analysis. *Methods Mol. Biol.* **721**, 153–171.

Meister, G., Landthaler, M., Dorsett, Y. and Tuschl, T. (2004). Sequence-specific inhibition of microRNA- and siRNA-induced RNA silencing. *RNA* **10**, 544–550.

Mica, E., Gianfranceschi, L. and Pe, M. E. (2006). Characterization of five microRNA families in maize. *J. Exp. Bot.* **57**, 2601–2612.

Miranda, K. C., Huynh, T., Tay, Y., Ang, Y. S., Tam, W. L., Thomson, A. M., Lim, B. and Rigoutsos, I. (2006). A pattern-based method for the identification of microRNA binding sites and their corresponding heteroduplexes. *Cell* **126**, 1203–1217.

Mochizuki, K. and Gorovsky, M. A. (2004). Conjugation-specific small RNAs in Tetrahymena have predicted properties of scan (scn) RNAs involved in genome rearrangement. *Genes Dev.* **18**(17), 2068–2073.

Mochizuki, K., Fine, N. A., Fujisawa, T. and Gorovsky, M. A. (2002). Analysis of a piwi-related gene implicates small RNAs in genome rearrangement in tetrahymena. *Cell* **110**(6), 689–699.

Morgulis, A., Coulouris, G., Raytselis, Y., Madden, T. L., Agarwala, R. and Schaffer, A. A. (2008). Database indexing for production MegaBLAST searches. *Bioinformatics* **24**, 1757–1764.

Nakamoto, M., Jin, P., O'Donnell, W. T. and Warren, S. T. (2005). Physiological identification of human transcripts translationally regulated by a specific microRNA. *Hum. Mol. Genet.* **14**, 3813–3821.

Nam, J. W., Kim, J., Kim, S. K. and Zhang, B. T. (2006). ProMiR II: a web server for the probabilistic prediction of clustered, nonclustered, conserved and nonconserved microRNAs. *Nucleic Acids Res.* **34**, W455–W458.

Nozawa, M., Miura, S. and Nei, M. (2010). Origins and evolution of microRNA genes in Drosophila species. *Genome Biol. Evol.* **2**, 180–189.

Nuovo, G., Lee, E. J., Lawler, S., Godlewski, J. and Schmittgen, T. (2009). In situ detection of mature microRNAs by labeled extension on ultramer templates. *Biotechniques* **46**, 115–126.

Obernosterer, G., Martinez, J. and Alenius, M. (2007). Locked nucleic acid-based in situ detection of microRNAs in mouse tissue sections. *Nat. Protoc.* **2**, 1508–1514.

O'Donnell, K. A., Wentzel, E. A., Zeller, K. I., Dang, C. V. and Mendell, J. T. (2005). c-Myc-regulated microRNAs modulate E2F1 expression. *Nature* **435**, 839–843.

Orom, U. A., Nielsen, F. C. and Lund, A. H. (2008). MicroRNA-10a binds the 5′UTR of ribosomal protein mRNAs and enhances their translation. *Mol. Cell* **30**, 460–471.

Oulas, A., Boutla, A., Gkirtzou, K., Reczko, M., Kalantidis, K. and Poirazi, P. (2009). Prediction of novel microRNA genes in cancer-associated genomic regions—a combined computational and experimental approach. *Nucleic Acids Res.* **37**(10), 3276–3287.

Pasquinelli, A. E., Reinhart, B. J., Slack, F., Martindale, M. Q., Kuroda, M. I., Maller, B., Hayward, D. C., Ball, E. E., Degnan, B., Muller, P., Spring, J., Srinivasan, A., *et al.* (2000). Conservation of the sequence and temporal expression of let-7 heterochronic regulatory RNA. *Nature* **408**, 86–89.

Qian, J., Zhang, Z., Liang, J., Ge, Q., Duan, X., Ma, F. and Li, F. (2011). The full-length transcripts and promoter analysis of intergenic microRNAs in Drosophila melanogaster. *Genomics* **97**, 294–303.

Ramkissoon, S. H., Mainwaring, L. A., Sloand, E. M., Young, N. S. and Kajigaya, S. (2006). Nonisotopic detection of microRNA using digoxigenin labeled RNA probes. *Mol. Cell. Probes* **20**, 1–4.

Reinhart, B. J., Slack, F. J., Basson, M., Pasquinelli, A. E., Bettinger, J. C., Rougvie, A. E., Horvitz, H. R. and Ruvkun, G. (2000). The 21-nucleotide let-7 RNA regulates developmental timing in Caenorhabditis elegans. *Nature* **403**, 901–906.

Ritchie, W., Theodule, F. X. and Gautheret, D. (2008). Mireval: a web tool for simple microRNA prediction in genome sequences. *Bioinformatics* **24**(11), 1394–1396.

Ronen, R., Gan, I., Modai, S., Sukacheov, A., Dror, G., Halperin, E. and Shomron, N. (2010). miRNAkey: a software for microRNA deep sequencing analysis. *Bioinformatics* **26**(20), 2615–2616.

Ru, S., Shen, W., Yang, P., Zhao, P. and Shao, Q. (2007). MicroRNA target prediction algorithm: present and future. *Chin. Bull. Life Sci.* **19**, 562–567.

Rusinov, V., Baev, V., Minkov, I. N. and Tabler, M. (2005). MicroInspector: a web tool for detection of miRNA binding sites in an RNA sequence. *Nucleic Acids Res.* **33**, W696–700.

Saba, R. and Booth, S. A. (2006). Target labelling for the detection and profiling of microRNAs expressed in CNS tissue using microarrays. *BMC Biotechnol.* **6**, 47.

Saba, R., Goodman, C. D., Huzarewich, R. L., Robertson, C. and Booth, S. A. (2008). A miRNA signature of prion induced neurodegeneration. *PLoS One* **3**, e3652.

Saetrom, O., Snove, O. Jr. and Saetrom, P. (2005). Weighted sequence motifs as an improved seeding step in microRNA target prediction algorithms. *RNA* **11**, 995–1003.

Sai Lakshmi, S. and Agrawal, S. (2008). piRNABank: a web resource on classified and clustered Piwi-interacting RNAs. *Nucleic Acids Res.* **36**, D173–177.

Sato, F., Tsuchiya, S., Terasawa, K. and Tsujimoto, G. (2009). Intra-platform repeatability and inter-platform comparability of microRNA microarray technology. *PLoS One* **4**, e5540.

Science (2001). Breakthrough of the year. Breakdowns of 2001. *Science* **294**, 2446.
Science (2002). Breakthrough of the year. The runners-up. *Science* **298**, 2297–2303.
Science (2003). Breakthrough of the year. Scorecard 2002. *Science* **302**, 2041.
Shi, R. and Chiang, V. L. (2005). Facile means for quantifying microRNA expression by real-time PCR. *Biotechniques* **39**, 519–525.
Shpiz, S., Kwon, D., Rozovsky, Y. and Kalmykova, A. (2009). rasiRNA pathway controls antisense expression of Drosophila telomeric retrotransposons in the nucleus. *Nucleic Acids Res.* **37**, 268–278.
Silahtaroglu, A. N. (2010). LNA-FISH for detection of microRNAs in frozen sections. *Methods Mol. Biol.* **659**, 165–171.
Silahtaroglu, A. N., Nolting, D., Dyrskjot, L., Berezikov, E., Moller, M., Tommerup, N. and Kauppinen, S. (2007). Detection of microRNAs in frozen tissue sections by fluorescence in situ hybridization using locked nucleic acid probes and tyramide signal amplification. *Nat. Protoc.* **2**, 2520–2528.
Smalheiser, N. R. (2003). EST analyses predict the existence of a population of chimeric microRNA precursor-mRNA transcripts expressed in normal human and mouse tissues. *Genome Biol.* **4**, 403.
Sperling, E. A., Vinther, J., Moy, V. N., Wheeler, B. M., Semon, M., Briggs, D. E. and Peterson, K. J. (2009). MicroRNAs resolve an apparent conflict between annelid systematics and their fossil record. *Proc. Biol. Sci.* **276**, 4315–4322.
Stark, A., Lin, M. F., Kheradpour, P., Pedersen, J. S., Parts, L., Carlson, J. W., Crosby, M. A., Rasmussen, M. D., Roy, S., Deoras, A. N., Ruby, J. G., Brennecke, J., *et al.* (2007). Discovery of functional elements in 12 Drosophila genomes using evolutionary signatures. *Nature* **450**, 219–232.
Taft, R. J., Glazov, E. A., Cloonan, N., Simons, C., Stephen, S., Faulkner, G. J., Lassmann, T., Forrest, A. R. R., Grimmond, S. M., Schroder, K., Irvine, K., Arakawa, T., *et al.* (2009). Tiny RNAs associated with transcription start sites in animals. *Nat. Genet.* **41**(5), 572–578.
Tan, L. P., Seinen, E., Duns, G., de Jong, D., Sibon, O. C., Poppema, S., Kroesen, B. J., Kok, K. and van den Berg, A. (2009). A high throughput experimental approach to identify miRNA targets in human cells. *Nucleic Acids Res.* **37**, e137.
Tanzer, A. and Stadler, P. F. (2004). Molecular evolution of a microRNA cluster. *J. Mol. Biol.* **339**, 327–335.
Terai, G., Komori, T., Asai, K. and Kin, T. (2007). miRRim: A novel system to find conserved miRNAs with high sensitivity and specificity. *RNA* **13**(12), 2081–2090.
Tran, T. T., Zhou, F., Marshburn, S., Stead, M., Kushner, S. R. and Xu, Y. (2009). De novo computational prediction of non-coding RNA genes in prokaryotic genomes. *Bioinformatics* **25**, 2897–2905.
Tweedie, S., Ashburner, M., Falls, K., Leyland, P., McQuilton, P., Marygold, S., Millburn, G., Osumi-Sutherland, D., Schroeder, A., Seal, R. and Zhang, H. (2009). FlyBase: enhancing Drosophila Gene Ontology annotations. *Nucleic Acids Res.* **37**, D555–559.
Tyagi, S., Vaz, C., Gupta, V., Bhatia, R., Maheshwari, S., Srinivasan, A. and Bhattacharya, A. (2008). CID-miRNA: a web server for prediction of novel miRNA precursors in human genome. *Biochem. Biophys. Res. Commun.* **372**(4), 831–834.
Udvardi, M. K., Czechowski, T. and Scheible, W. R. (2008). Eleven golden rules of quantitative RT-PCR. *Plant Cell* **20**, 1736–1737.
Vagin, V. V., Sigova, A., Li, C., Seitz, H., Gvozdev, V. and Zamore, P. D. (2006). A distinct small RNA pathway silences selfish genetic elements in the germline. *Science* **313**, 320–324.

Valoczi, A., Hornyik, C., Varga, N., Burgyan, J., Kauppinen, S. and Havelda, Z. (2004). Sensitive and specific detection of microRNAs by northern blot analysis using LNA-modified oligonucleotide probes. *Nucleic Acids Res.* **32**, e175.

Varallyay, E., Burgyan, J. and Havelda, Z. (2007). Detection of microRNAs by Northern blot analyses using LNA probes. *Methods* **43**, 140–145.

Varallyay, E., Burgyan, J. and Havelda, Z. (2008). MicroRNA detection by northern blotting using locked nucleic acid probes. *Nat. Protoc.* **3**, 190–196.

Vazquez, F., Vaucheret, H., Rajagopalan, R., Lepers, C., Gasciolli, V., Mallory, A. C., Hilbert, J. L., Bartel, D. P. and Crete, P. (2004). Endogenous trans-acting siRNAs regulate the accumulation of Arabidopsis mRNAs. *Mol. Cell* **16**(1), 69–79.

Wang, X., Zhang, J., Li, F., Gu, J., He, T., Zhang, X. and Li, Y. (2005). MicroRNA identification based on sequence and structure alignment. *Bioinformatics* **21**(18), 3610–3614.

Weaver, D. B., Anzola, J. M., Evans, J. D., Reid, J. G., Reese, J. T., Childs, K. L., Zdobnov, E. M., Samanta, M. P., Miller, J. and Elsik, C. G. (2007). Computational and transcriptional evidence for microRNAs in the honey bee genome. *Genome Biol.* **8**, R97.

Wei, Y., Chen, S., Yang, P., Ma, Z. and Kang, L. (2009). Characterization and comparative profiling of the small RNA transcriptomes in two phases of locust. *Genome Biol.* **10**, R6.

Wheeler, B. M., Heimberg, A. M., Moy, V. N., Sperling, E. A., Holstein, T. W., Heber, S. and Peterson, K. J. (2009). The deep evolution of metazoan microRNAs. *Evol. Dev.* **11**(1), 50–68.

Wilusz, J. E., Sunwoo, H. and Spector, D. L. (2009). Long noncoding RNAs: functional surprises from the RNA world. *Genes Dev.* **23**, 1494–1504.

Wu, Y., Wei, B., Liu, H., Li, T. and Rayner, S. (2011). MiRPara: a SVM-based software tool for prediction of most probable microRNA coding regions in genome scale sequences. *BMC Bioinformatics* **12**, 107.

Xie, Z., Allen, E., Fahlgren, N., Calamar, A., Givan, S. A. and Carrington, J. C. (2005). Expression of Arabidopsis MIRNA genes. *Plant Physiol.* **138**, 2145–2154.

Xiong, H., Qian, J., He, T. and Li, F. (2009). Independent transcription of miR-281 in the intron of ODA in Drosophila melanogaster. *Biochem. Biophys. Res. Commun.* **378**, 883–889.

Xue, C. and Li, F. (2008). Finding noncoding RNA transcripts from low abundance expressed sequence tags. *Cell Res.* **18**, 695–700.

Xue, C., Li, F., He, T., Liu, G. P., Li, Y. and Zhang, X. (2005). Classification of real and pseudo microRNA precursors using local structure-sequence features and support vector machine. *BMC Bioinformatics* **6**, 310.

Yang, J. H., Shao, P., Zhou, H., Chen, Y. Q. and Qu, L. H. (2010). deepBase: a database for deeply annotating and mining deep sequencing data. *Nucleic Acids Res.* **38**, D123–D130.

Yauk, C. L., Rowan-Carroll, A., Stead, J. D. and Williams, A. (2010). Cross-platform analysis of global microRNA expression technologies. *BMC Genomics* **11**, 330.

Yin, H. and Lin, H. (2007). An epigenetic activation role of Piwi and a Piwi-associated piRNA in Drosophila melanogaster. *Nature* **450**, 304–308.

Yousef, M., Nebozhyn, M., Shatkay, H., Kanterakis, S., Showe, L. C. and Showe, M. K. (2006). Combining multi-species genomic data for microRNA identification using a Naive Bayes classifier. *Bioinformatics* **22**(11), 1325–1334.

Yue, J., Sheng, Y. and Orwig, K. E. (2008). Identification of novel homologous microRNA genes in the rhesus macaque genome. *BMC Genomics* **9**, 8.

Zhang, B. H., Pan, X. P., Wang, Q. L., Cobb, G. P. and Anderson, T. A. (2005). Identification and characterization of new plant microRNAs using EST analysis. *Cell Res.* **15**, 336–360.

Zhang, Y., Zhou, X., Ge, X., Jiang, J., Li, M., Jia, S., Yang, X., Kan, Y., Miao, X., Zhao, G., Li, F. and Huang, Y. (2009). Insect-specific microRNA involved in the development of the silkworm Bombyx mori. *PLoS One* **4**, e4677.

Zhang, Y., Wang, X. and Kang, L. (2011). A k-mer scheme to predict piRNAs and characterize locust piRNAs. *Bioinformatics* **27**, 771–776.

Zhou, X., Duan, X., Qian, J. and Li, F. (2009). Abundant conserved microRNA target sites in the 5′-untranslated region and coding sequence. *Genetica* **137**, 159–164.

Zhu, E., Zhao, F., Xu, G., Hou, H., Zhou, L., Li, X., Sun, Z. and Wu, J. (2010). mirTools: microRNA profiling and discovery based on high-throughput sequencing. *Nucleic Acids Res.* **38**(Suppl.), W392–W397.

Zou, Q., Guo, M., Liu, Y. and Xuan, P. (2010). DuplexFinder: predicting the miRNA-miRNA* duplex from the animal precursors. *Int. J. Bioinform. Res. Appl.* **6**(1), 69–81.

Index

Note: Page numbers followed by "*f*" indicate figures, and "*t*" indicate tables.

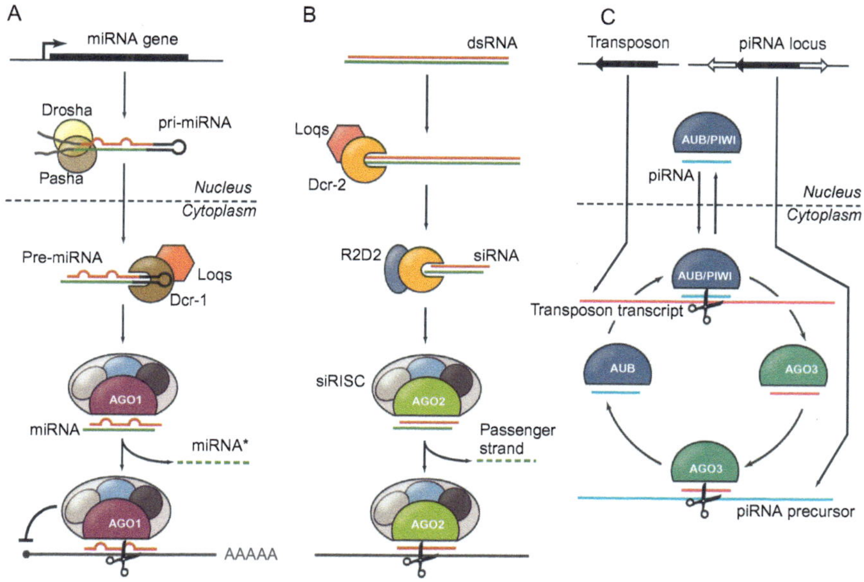

PLATE 1 Schematic representation of the small RNA pathways in *Drosophila*. (A) The miRNA pathway is initiated by the transcription of miRNA genes. Primary miRNA (pri-miRNA) transcripts are first processed by the Drosha/Pasha complex in the nucleus then exported to the cytoplasm as pre-miRNAs. There, Dcr-1 and loquacious (loqs) complete the processing and deliver the mature miRNA to AGO1-containing RISC. The miRNA* (star) is eliminated and the guide miRNA directs translational repression or cleavage of the cognate mRNA. (B) The siRNA pathway is initiated by double-stranded RNA of viral or genomic origin which is recognised and cleaved by Dcr-2 with the help of loqs. The resulting double-stranded siRNAs are delivered to AGO2-containing RISC by Dcr-2 and R2D2. The passenger strand is eliminated and the guide siRNA directs the degradation of the target RNA via AGO2 catalytic activity. (C) The piRNA pathway is initiated by maternally deposited piRNAs loaded into Aubergine (Aub)/Piwi proteins. Through an amplification loop that involves sense and antisense transcripts, AGO3 and Aub, the piRNA pool is amplified. The newly produced piRNAs loaded into Aub are transported to the nucleus where they are thought to be involved in chromatin modifications. (For b/w version, see page 7 in the volume.)

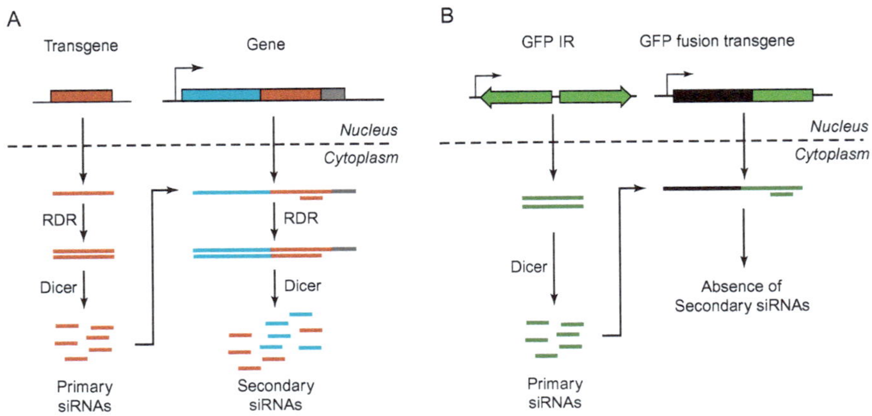

PLATE 2 Examples of transitivity assays in plants (A) and *Drosophila* (B). (A) In plants, aberrant transcripts produced from an integrated transgene corresponding to a fragment of an endogenous gene are converted into dsRNA by an RNA-dependent RNA polymerase (RDR) and cleaved by Dicer. The resulting siRNAs (red) are used to prime the RDR-dependent synthesis of a dsRNA molecule using the homologous endogenous gene transcript as template. Consequently, the newly synthesised dsRNA molecule contains sequence of the surrounding transcript (blue). The presence of secondary siRNAs corresponding to the surrounding regions (blue) is a signature of transitivity. (B) A similar assay was designed in *Drosophila* but failed to identify secondary siRNAs. An inverted repeat (IR) directed against the GFP (green) was expressed from the fly genome. GFP siRNAs were detected (green). However, secondary siRNAs corresponding to the surrounding sequences (black) of a GFP fusion transcript were not detected arguing against transitivity in *Drosophila*. (For b/w version, see page 14 in the volume.)

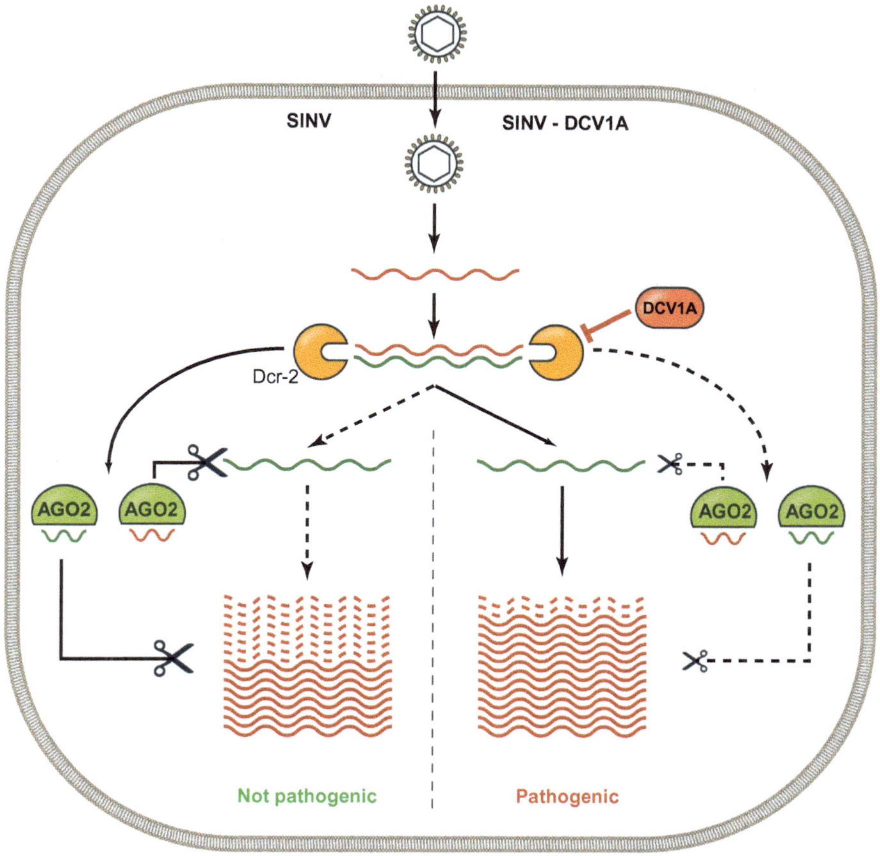

PLATE 3 Model illustrating the impact of VSR on RNAi, viral replication and outcome as exemplified by the SINV (left) and DCV1A-expressing SINV (right). Upon SINV infection (left side), the virus releases its positive strand genome (red) which is first translated and then used as a template for synthesis of the negative strand genome (green). In turn, the negative strand RNA serves as template for the synthesis of numerous positive strand RNAs that will be translated to increase the pool of viral proteins and encapsidated in newly synthesised viruses. Through this replication process, RNAi targets various molecular forms of the viral RNA. First, dsRNA replication intermediates (green and red) are cleaved by Dcr-2 and SINV-derived siRNAs are loaded into AGO2. These siRNAs guide the cleavage of viral RNAs of both polarity, limiting the accumulation of viral RNAs. Under these wild-type conditions, SINV is not pathogenic for insects. (Right side) The expression of the long dsRNA-specific binding protein DCV1A from the SINV genome protects the dsRNA intermediates of replication from degradation by Dcr-2. Consequently, fewer siRNAs are produced and loaded into AGO2, leading to a greater accumulation of viral RNAs. Under these conditions, DCV1A-expressing SINV is pathogenic for *Drosophila* contrary to the wild-type virus. (For b/w version, see page 22 in the volume.)

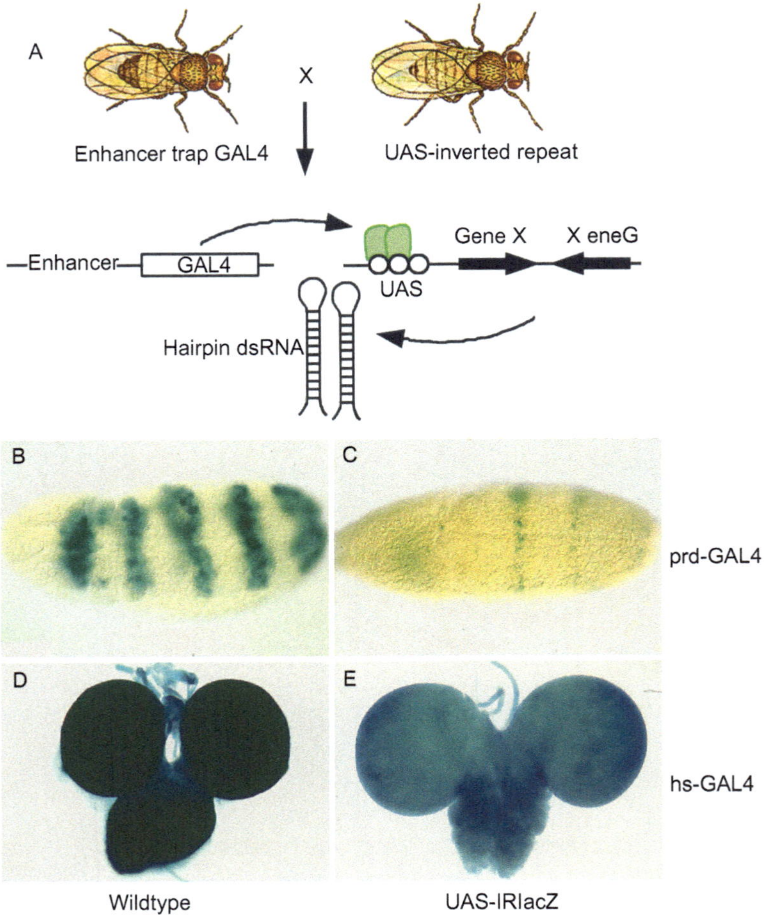

PLATE 4 *Drosophila melanogaster* RNAi transgene system. (A) Strategy for generation of transgenic RNAi. A fragment of a gene is dimerized to produce a long double-stranded hairpin RNA. The transgene is under the control of a generic promoter containing the GAL4-responsive upstream activator sequence (UAS) element in an inducible expression vector (right fly). Upon mating UAS-*dsRNA* transgenic flies with flies carrying a tissue- or cell-specific GAL4 transgene (left fly), the F1 progeny generated a hairpin-like dsRNA in all cells that expressed the GAL4 gene and in these cells the target genes were silenced (Kennerdell and Carthew, 2000). (B and D) F1 progeny carrying a GAL4 transgene and reporter UAS-*lacZ* gene but no UAS-*IRlacZ* transgene. (C and E) F1 progeny carrying the UAS-*IRlacZ* that was predicted to make lacZ dsRNA. (B and C) Stage 10 embryos carrying paired-GAL4 showing a pair-rule pattern of lacZ expression that was inhibited by the presence of UAS-IRlacZ. (D and E) Brain complex and thoracic nerve cord from late third-instar hsp70-GAL4 larvae after heat shock. Photos adapted from Kennerdell and Carthew (2000). (For b/w version, see page 85 in the volume.)

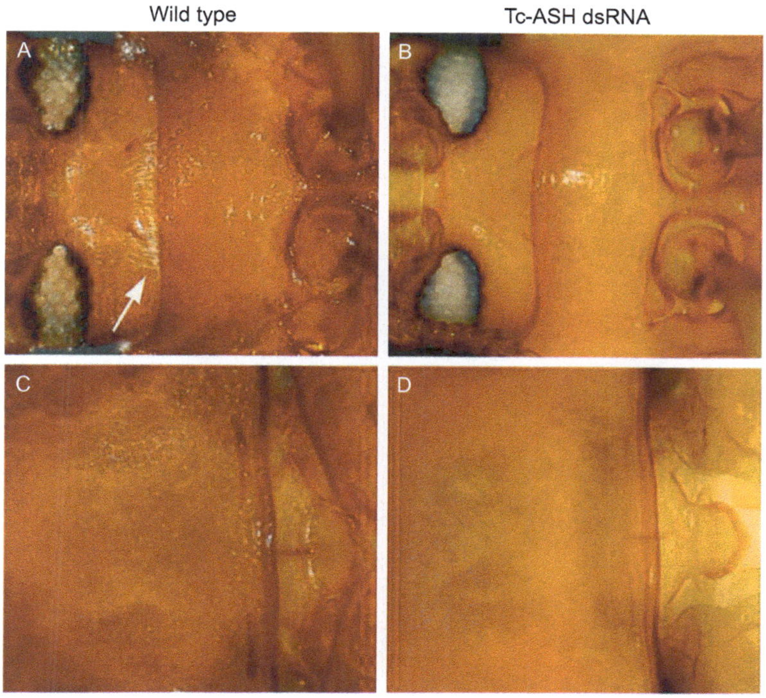

PLATE 5 *Tribolium castaneum* systemic larval RNAi revealed by *Tc-ASH* RNAi phenotypes. Non-injected *pu11* adult (A, C) and a *pu11* adult in which *Tc-ASH* dsRNA was injected during the last larval stage (B, D). Ventral view of the head and first thoracic sternum (A, B) and dorsal view of pro- and mesonotum (C, D). The entire wild-type adult cuticle is covered by small indentations, each associated with a sensory bristle (A, C). There is also a fringe of large bristles between the head and thorax (arrow in A). In *Tc-ASH* dsRNA-injected beetles, the cuticle was completely smooth (B, D) and had few sensory structures. The large bristles between head and thorax were also lost. (Photos reproduced courtesy of Yoshinori Tomoyasu and Robin E. Denell). (For b/w version, see page 88 in the volume.)

PLATE 6 dsRNA transgenic plants were less damaged. (A) *dsCYP6AE14* transgenic cotton plants were less damaged by bollworms. To test the protection of cotton bolls by dsRNA transgenic cotton, *Helicoverpa armigera* larvae that had previously been reared on leaves of the wild type and *dsCYP6AE14* plants for 10 day were transferred to bolls (about 9 days post-anthesis). After 1 day, the larvae from wild type leaves burrowed into bolls and consumed the contents (upper panel), whereas the larvae from *dsCYP6AE14* leaves chewed only shallow gouges in the boll surface, and the boll contents were almost intact (lower panel). These data indicate that the transgenic cotton plants expressing the insect P450 dsRNA were less damaged by cotton bollworms in comparison with the untransformed control. Photos adapted from (Mao *et al.*, 2011). (B) F1 corn plants expressing a V-ATPase A dsRNA were protected from WCR feeding damage. To test whether corn plants expressing a WCR-derived dsRNA were protected from rootworm feeding damage, Baum *et al.* (2007) transformed a V-ATPase A dsRNA expression cassette into corn plants. The maize transformants were backcrossed with an inbred line to generate F1 hybrid progeny. F1 hybrid plants were evaluated using a growth chamber assay to assess rootworm feeding damage. The plant on left is a non-transgenic control with average root damage, whereas the dsRNA transgenic plant on the right shows obvious root protection. (Photos reproduced courtesy of James Baum and James Roberts). (For b/w version, see page 99 in the volume.)

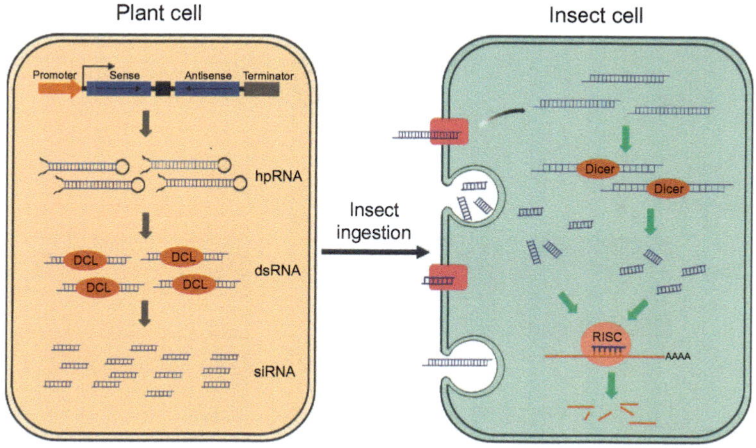

PLATE 7 Insect RNAi mediated by transgenic plants. Transgenic plants express hairpin RNAs (hpRNA) directed against an insect gene. HpRNAs are diced into the intermediate product of dsRNAs, and then processed into siRNAs by plant Dicer enzymes. The dsRNAs and/or the siRNAs are orally delivered into larvae feeding on transgenic plants by SID-1 homologue-mediated transmembrane channel, scavenger receptor-mediated endocytosis or other pathways. The siRNAs, ingested directly from the plants or processed by insect Dicers after ingestion of long dsRNAs, elicit cleavage of the target mRNA in RISC in midgut cells. (For b/w version, see page 103 in the volume.)

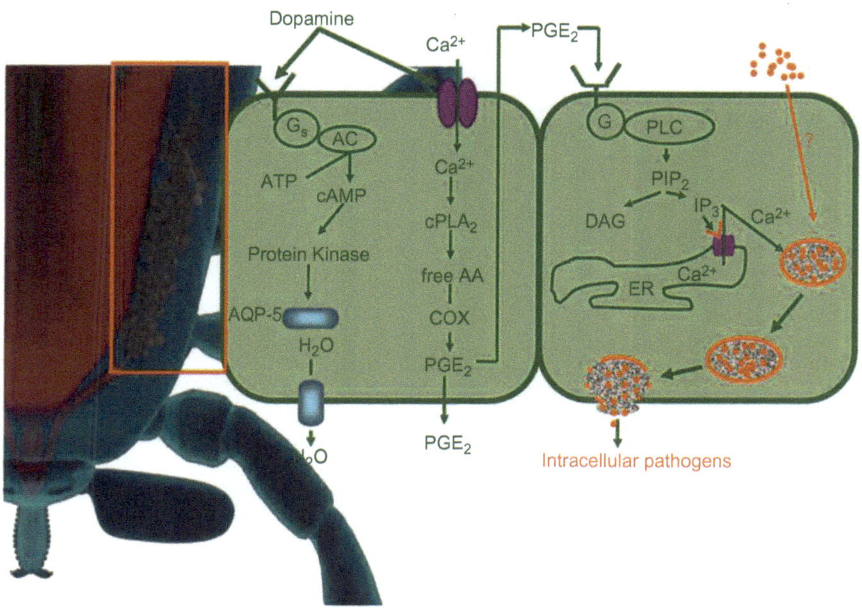

PLATE 8 The known and proposed cascade of reactions controlling prostaglandin E_2 synthesis and tick salivary secretions (Bowman and Sauer, 2004). Dopamine released from nerves stimulates fluid secretion through an increase in cAMP and activation of a water channel protein. Dopamine also stimulates the influx of Ca^{2+}, activating an intracellular PLA_2 and subsequently releasing arachidonic acid (AA) which is converted to PGE_2 by a cyclooxygenase pathway. PGE_2 is either secreted into the saliva or functions as a local hormone by interacting with PGE_2 receptors to increase IP3. IP3 mobilizes intracellular Ca^{2+} which is critical for regulating exocytosis of salivary proteins. Tick-transmitted pathogens are thought to exploit transcytotic pathways, that is they are taken in by endocytosis and released by exocytosis (modified from Bowman and Sauer, 2004). (For b/w version, see page 122 in the volume.)

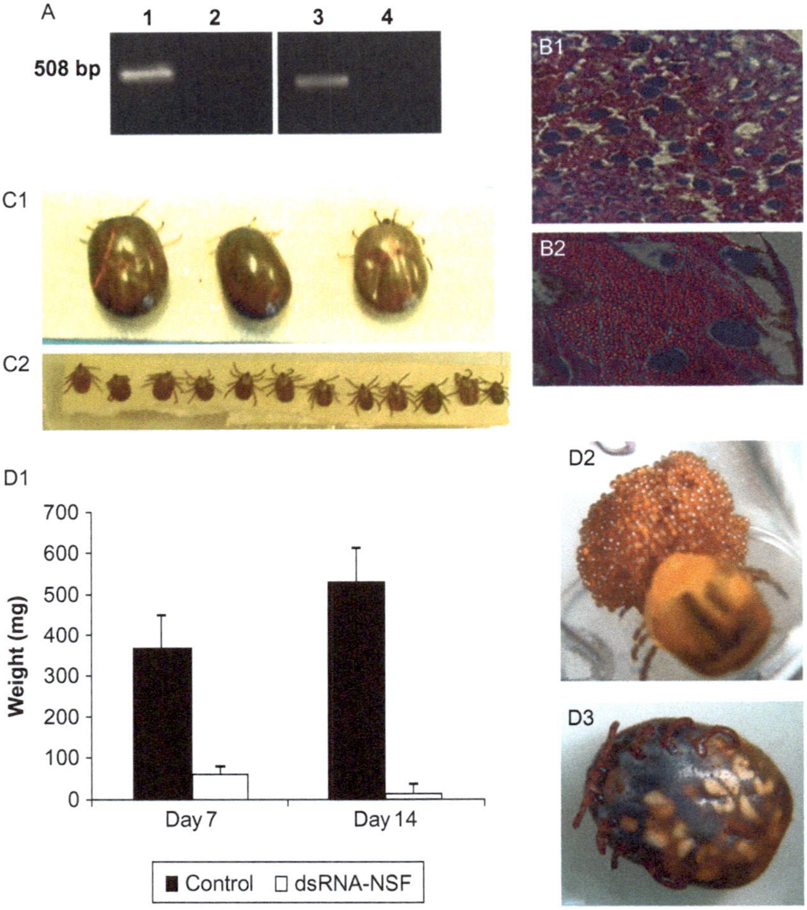

PLATE 9 *In vivo* RNAi in ticks. The *N*-ethylmaleimide-sensitive fusion factor (NSF) dsRNA and an irrelevant GFP-dsRNA delivered by microinjection in unfed female *Dermacentor variabilis*. After microinjections of dsRNA, ticks are left overnight in an incubator and infested on a rabbit. (A) Transcriptional expression of NSF gene in partially fed tick tissues, (1,3) control midgut and salivary gland tissues, (2,4) NSF-dsRNA-injected midgut and salivary gland tissues. (B1 and B2) Thick sections of control and NSF-dsRNA salivary glands stained with dye demonstrating the disruption of secretory vesicles. (C1 and C2) Tick engorgement in control and NSF-dsRNA-treated ticks after 15 days of attachment on rabbit. (D1) Average engorgement weight of ticks on day 7 and 14 among control and treated tick groups. (D2 and D3) Oviposition in control and dsRNA-NSF injected ticks. (For b/w version, see page 138 in the volume.)